The Politics of Bioethics

Routledge Studies in Science, Technology and Society

The Politics of Bioethics

Alan Petersen

Routledge
Taylor & Francis Group
New York   London

First published 2011
by Routledge
711 Third Ave, New York, NY 10017

Simultaneously published in the UK
by Routledge
2 Park Square, Milton Park, Abingdon, Oxon OX14 4RN

Routledge is an imprint of the Taylor & Francis Group, an informa business

© 2011 Taylor & Francis

First issued in paperback 2013

The right of Alan Petersen to be identified as author of this work has been asserted in accordance with sections 77 and 78 of the Copyright, Designs and Patents Act 1988.

Typeset in Sabon by IBT Global.

Library of Congress Cataloging in Publication Data

Petersen, Alan R., Ph. D.
  The politics of bioethics / by Alan Petersen.
    p. ; cm.—(Routledge studies in science, technology and society ; 14)
  Includes bibliographical references and index.
  1. Bioethics—Political aspects. 2. Medical ethics—Political aspects. I. Title.
  II. Series: Routledge studies in science, technology, and society ; 14.
  [DNLM: 1. Bioethics. 2. Bioethical Issues. 3. Politics. WB 60]
QH332.P48 2011
174'.957—dc22
2010035404

ISBN13: 978-0-415-99006-6 (hbk)
ISBN13: 978-0-415-85139-8 (pbk)

Contents

Acknowledgments

This book has benefited from the assistance offered by many individuals over a number of years. I am especially grateful to the following people who have generously offered references, insights, and items of information: Oonagh Corrigan, Iain Wilkinson, Erica Haimes, Michael Barr, Steven Wainwright, Clare Williams, Alison Anderson, Helen Wallace, Cathy Waldby, Jessica Hutchings, Jim Macnamara, Mark Stranger, Loane Skene, Craig Cormick, Justin Oakley, Rob Sparrow, Barbara Katz Rothman, Diana Bowman, and Kate Seear. Some of the chapters draw on or develop ideas published elsewhere. A number of the ideas in Chapter 2 build on arguments outlined in an article published in *Monash Bioethics Review* (v.28, n.1). Chapter 3 draws on some work originally published in *Sociology of Health and Illness* (v.27, n.2) and *Genomics, Society and Policy* (v.3, n.1). Chapter 5 develops ideas and draws on some material from articles published in *Health, Risk and Society* (v.1, n.3) and *New Formations* (v.60), respectively. Chapter 6, while completely new, develops ideas that arise from various projects undertaken in collaboration with Alison Anderson, Stuart Allen, and Clare Wilkinson. The usual disclaimers apply, of course. I would like to thank Benjamin Holtzman and Max Novick, at Taylor and Francis, for their assistance and patience for the long delay in the completion of the manuscript. Finally, as always, I am greatly indebted to Ros Porter, without whose love and support this book would not have been completed.

1 Bioethics as Politics

If recent news reports and science commentaries can be believed, biotechnology will soon profoundly alter our lives. Hardly a week goes by without the announcement of some new biomedical 'breakthrough' that will deliver new treatments that will reduce people's suffering, enhance their physical performance or mental abilities, or increase their lifespan. Stem cell treatments, genetic therapies, and nanotechnologies, along with other innovations, it is claimed, have the potential to 'revolutionise' healthcare and deliver significant economic benefits in the years ahead. According to many scholars, biotechnology will change our very conceptions of the natural and the normal and unsettle our assumptions about what it means to live a 'good', 'healthy' life. The rapid development of biotechnologies and the wide range of purported applications have heightened concerns about where technologies are leading and prompted some to ask whether they present dangers that outweigh the benefits. Biotechnology, it is contended, poses questions that go beyond the familiar ones of safety and equality of access to innovations to encompass those about the kind of human being and the sort of society that will be created in the future (Kass, 2003: xvi).

In their endeavours to make sense of and find guidance about how best to respond to the challenges posed by biotechnologies, many authorities and citizens have looked to bioethics. Over the last three decades, and especially during the last decade, bioethics has grown rapidly as a professional field and has gained increasing salience in public and policy debates. More and more, bioethics is called upon to provide moral direction in relation to the various quandaries posed by particular biotechnology innovations. Bioethicists have been asked to help adjudicate on a range of matters such as when life begins and ends, how best to protect the autonomy and rights of the individual, how to prevent discrimination arising from the use of new technologies, and how to achieve justice in the allocation of scarce healthcare resources. However, in recent years, bioethics has been subject to a growing number of criticisms from scholars working in various fields, with concerns raised about its limitations and implications in practice, and calls to develop approaches that move 'beyond bioethics' to meet the profound challenges posed by the biosciences and biotechnologies. Critics have wondered not only whether *too much* is expected of bioethics to help resolve problems posed by biotechnologies but also whether bioethics presents its own dangers.

This book explores the politics of bioethics. It examines the largely unintended consequences of the application of bioethics' concepts and principles and modes of analysis to help resolve perceived problems associated with the development of new and emerging biotechnologies. Making reference to some biotechnologies that are widely predicted to have significant impacts on health, healthcare, and society in the future, I ask, what are the implications of how bioethics ideas are employed *in practice* for how scientists, policymakers, other influential groups (e.g. business), and lay communities respond to these technologies? In the chapters, I emphasise the significant, under-acknowledged role that *expectations* play in shaping biotechnologies and the attendant social responses, and I examine the interests, values, and ideologies that contribute to creating and sustaining these expectations. Bioethics concepts and principles, I argue, have served as a tool of *governance*, in helping to engender consent and legitimacy for the development of technologies that involve many uncertainties, including the nature and timing of specific applications, the benefits, the biophysical risks and other dangers (e.g. new pernicious forms of surveillance, the reinforcement of inequalities), and the social responses. In developing this analysis, I aim to advance debate about the socio-political implications of bioethics and help lay the groundwork for developing new critical perspectives on the biosciences and biotechnologies.

WHAT IS DIFFERENT ABOUT THIS BOOK?

There are many books on bioethics, and one may rightly ask, what is different about this one? To clarify its focus, I should point out at the outset what this book is *not* about. Firstly, the book does not seek to explain the origins and development of bioethics as a discipline or field of practice. Indeed, it steers away from any attempt to describe and account for the rise of bioethics. For those who are interested, a number of books, chapters, and articles seek to do just this (e.g. Fox and Swazey, 2008; Jonsen, 1998; Rothman, 1991). I do refer in passing to histories of bioethics, mainly in this chapter, but I do this in order to underline how histories to date have served to constrain rather than enlighten thinking about this field and its impacts. Secondly, I do not aim to describe what bioethics *is* or what bioethicists *do*. Again, there are many publications, particularly those oriented to students and practitioners of the health sciences and philosophy, which examine the nature and scope of the field of bioethics and the work of bioethicists (e.g. Kuhse and Singer, 2006; Oakley, 2009; Singer and Viens, 2008). However, I am interested in how bioethics is *represented* and how certain ideas, concepts, principles, and modes of analysis and reasoning associated with this field are used to advance particular goals. As I explain in the next section, 'bioethics' is not a stable, easily definable field; it is subject to contestation and to utilisation in various ways in different contexts and at different times. Moreover, 'bioethicists' constitute a diverse and shifting identity category, and it often proves difficult to identify those to whom this label applies in particular situations. This allows much scope for this identity label to be adopted by those who seek to advance certain agenda. Thirdly, I should make clear that in making critical observations on the *uses* of bioethics' ideas, I do not wish to contribute to the 'boundary disputes' that have characterised sociological engagements with bioethics in recent years which, in my view, ultimately proves fruitless and has served to create entrenched positions. (For an example, see Turner, 2009.) Rather, my aim is to offer a *sociological* perspective on the discourses and practices of bioethics in order to help chart a new direction for the critical normative analysis of the biosciences and biotechnologies in the future.

I offer what might be broadly described as a social construction of knowledge perspective. This encompasses a range of different approaches to knowledge; to understanding how we know what we know. However, all tend to share the assumption that knowledge is always 'socially embedded'; that is, it is a product of and contingent on a particular set of historical and socio-cultural conditions and is therefore subject to change. Those who adopt this broad perspective challenge the claim that one can know something absolutely, without doubt, once and for all. This approach sits uneasily in a science-based culture because science tends to be presented as objective, independent of context, cumulative, and oriented to uncovering 'the truth' of the phenomenon that is studied. It unsettles the truth claims of those who subscribe to objectivism. It is therefore radical in its implications. More specifically, I am interested in the political implications of bioethics' particular ways of knowing and the consequences for social action. Consequently, my approach may be described more specifically as a *politics of knowledge* perspective. Later in the chapter, I discuss in more detail the concepts and theories that shape my approach. In brief, I am concerned with the ways in which bioethics ideas are employed to justify certain lines of action or to legitimise inaction, and the related expertise that is utilised in the effort to realise desired outcomes. More than this, I am interested in how bioethics ideas may help reinforce dominant relations of power and detrimentally affect or potentially affect people's health and wellbeing by drawing attention away from substantive questions, such as whether particular technologies should be developed at all.

In this chapter, I outline my guiding assumptions, in relation to the operations of politics and power in modern societies, drawing on a range of ideas from sociology and science and technology studies. Through this enquiry, my aim is to encourage researchers and students in the biosciences and biotechnologies and those who are interested in understanding the broad implications of these fields to reflect on the kinds of knowledge and practices that are needed to address the challenges posed by the biosciences and biotechnologies in the future.

DEBATE ABOUT THE CONTENT OF BIOETHICS

The question of what constitutes, or should constitute, 'bioethics' is one of ongoing academic debate. Like many fields of professional or quasi-professional practice, in this field there are contending definitions about the content and boundaries of knowledge and practice. One then needs to be careful in making sweeping claims about 'bioethics' and its implications. One of the difficulties in delineating bioethics is that it traverses a range of areas that are variously known as 'medical ethics', 'clinical ethics', 'research ethics', and 'biomedical ethics' (Hedgecoe, 2004a: 122). Some of what is called 'medical ethics' or 'clinical ethics' focuses mostly on the norms surrounding the micro-dynamics of medical decision-making and patient–doctor relations rather than on the normative and justice implications of new and emergent technologies like stem cell technologies, cloning, genetic testing, and so on (see, e.g. Komesaroff, 2008). To further complicate the picture, in some medical schools in the US and UK at least, 'ethics' is taught within programs called the medical humanities (Fox and Swazey, 2008: 39–41), an interdisciplinary field which is itself subject to contestation (Petersen, et al., 2008).

As De Vries, et al., point out, bioethics is not 'a monolithic entity, with a single perspective and mode of enquiry, reinforced by a cadre of leaders whose position and expertise are unchallenged—an orthodox professional group capable of enforcing such tight discipline that the "field" speaks with one voice on all issues' (2007: 2; see also De Vries, 2007). It constitutes a diverse array of activities, which 'range from serving on national commissions, to providing advice to the pharmaceutical industry, to consulting at the bedside when conflict between and among caregivers, family members and patients has stalled decision-making, to debating in journals the wisdom that one's colleagues have shown in their articles, to acting as public intellectuals when the instant op-ed piece is needed or when CNN calls' (De Vries, et al., 2007: 2). Although in this book my primary interest is in how bioethics' ideas are used *in practice* in relation to new and emergent technologies that are widely expected to have substantial social impacts, rather than the character of bioethics debates in journals and other forums, I do examine some of these debates for what they reveal about the preoccupations of those who work in the field of bioethics or who utilise bioethics concepts in analysing issues. (See, for example, Chapters 4 and 6.)

'Bioethicist' is an equally plural entity, constituting a diverse array of actors, including lawyers, physicians, theologians, philosophers, anthropologists, sociologists, historians, and patient activists, among others. This diversity of expertise and fields of interest makes it difficult to invalidate any group's claim to 'bioethics' expertise or to prevent anyone using the label 'bioethicist' (De Vries, et al., 2007: 2). However, as in other avowedly multi-disciplinary or interdisciplinary fields, such as public health, in practice there tends to be an assumed hierarchy of expertise, with some disciplines or specialist fields of practice enjoying more status and legitimacy than others. For instance, the views of philosophers (particularly moral philosophers), lawyers, theologians, and medical practitioners tend to carry more weight in bioethics debates than the views of sociologists, anthropologists, and historians. Further, bioethics committees have been found to be dominated by experts from medical science (particularly medical genetics) and law rather than by those who identify themselves as bioethicists (see, e.g. Salter and Salter, 2007: 567).

The question of what constitutes bioethics and who claims to be a bioethicist are not just matters of 'academic' interest. In a society where expert knowledge is accorded considerable status and where the possession of particular expertise confers power in the ability to define 'truth' within a field, understanding the basis of claims to knowledge is integral to understanding the operations of power and who 'wins' and 'loses', or is likely to, from decisions that affect health and wellbeing. Competing claims makers struggle to delineate the parameters of and colonise the field and seek to impose their own preferred definitions of the situation. Like most areas of expertise, bioethics knowledge and practice is sustained through 'boundary work' involving the policing of the field's borders and ensuring that only those who are deemed as qualified to speak about and act upon particular issues do so (Gieryn, 1999).

The politics of bioethics operates at a number of different levels, which will be explored in the book. These include the ability to claim authority over and delineate the boundaries of the field and thereby affect the capacity of other stakeholders or potential stakeholders to shape responses to the biosciences and biotechnologies. Bioethics serves to guide action in an important way through 'limiting the scope of actual decisionmaking to "safe" issues', or what has been called 'non-decision making' (Bachrach and Baratz, 1963: 632)—and by *narrowing* the scope for debate by referring to a certain range of established procedures and/or abstract, universal principles—so-called 'principlism' (Evans, 2002: 87–93). Principalism, which is widely seen to have been pioneered by Tom Beauchamp and James Childress, has had a profound influence on bioethics. In their multi-edition book, *Principles of Biomedical Ethics* (2001), originally published in 1979, Beauchamp and Childress proposed four midlevel moral principles that were to serve as essential guides to bioethical deliberations, namely, autonomy, beneficence, nonmaleficence, and justice (Walker, 2009: 8).

Although principlism has been subject to much critique and defence (including rejoinders by Beauchamp and Childress, who claim that it was never intended to be used in a deductive way, where principles are merely applied to specific cases (Wolf, 1996: 16)), this approach continues to hold sway in 'mainstream' bioethics. In commenting on the influence of principlism, Walker observes:

> Much current mainstream bioethical discussion, in both academic and institutional contexts, continues to frame questions and answers in terms of these guiding principles; they have achieved a wide resonance, if not a strictly uniform conception or priority ordering. . . . The values enshrined in the four cardinal principles are foundational for bioethics and have been formative for it not only intellectually but professionally.
> (Walker, 2009: 8–9)

Along with other writers (e.g. Evans, 2002; Fox and Swazey, 2008), I argue that the appeal of principlism can be explained by its congruence with dominant values, especially those pertaining in the United States , where its tenets arose. Specifically, it accords with the ideology of liberal individualism—a fact that accounts in large measure for its wide appeal in societies that are dominated by neo-liberal rule. Principlism is a product of a particular time and historical context and has provided bioethics with a common language and method for conceptualising problems (Wolpe, 1998: 55). The use of a specialised language, to create a shared bioethics community and exclude outsiders, is an important aspect of an emerging discipline and its work of boundary maintenance. However, as Wolpe observes, in the future, new categories will be formulated that will make the language of principlism irrelevant; indeed this reformulation is already occurring (1998: 55).

The deployment within bioethics of particular academic disciplines, especially moral philosophy, and perspectives (particularly utilitarianism) to research and to debate the nature of problems has lent credibility to the field by professionalizing practice and establishing the authority of expert practitioners. Bioethics' 'top-down' prescriptions, however, deny the diversity of perspectives on the biosciences and emergent biotechnologies and conceal their gender, class, and ethnicity biases. Increasingly, 'mainstream' bioethics concepts and principles have been applied globally to new sites and societies in what are arguably 'over-extended' and inappropriate ways. In the view of Salter and Salter (2007: 555), 'Bioethics has become the political means for the creation of a global moral economy where the trading and exchange of ideas is normalized and legitimated'. One of the aims of this book is to reveal the various manifestations and implications of bioethics' knowledge, as it has been applied within and across societies. In particular, I explore the dangerous implications of what might be described as 'bioethics imperialism'—a term that underscores the potentially damaging impacts of the extension of its rich Western-, particularly US-focused approach to ever growing spheres of life.

Imperialism may seem a strong word to describe the international diffusion and growing application of bioethics concepts, principles, and related practices. However, if one conceives of imperialism as 'a system in which a country rules other countries, sometimes having used force to get power over them' (Cambridge Advanced Learner's Dictionary, 2010), then this description may not seem too far-fetched. Although direct force may not be used, the assumption that a predominantly US view of the world reflected in bioethics should provide the basis for deliberating on fundamental issues of life has come to operate as a form of hegemony or cultural domination. No force is required because there is almost universal consensus that bioethics can and *should* provide direction on such issues. To help loosen the powerful hold of bioethics on thinking and to help set the scene for the chapters that follow, I believe it is helpful to examine an important way in which the field is legitimised, namely, the representations of its *histories*.

HISTORIES OF BIOETHICS

Many histories have been written about bioethics, particularly in the United States, where the institutionalisation of bioethics is claimed to have first occurred. As Fox and Swazey (2008) note, in their recent account of the rise of bioethics in the US, there are many narrative accounts to explain the rise of bioethics. As they observe,

> Bioethics in the United States has been noticeably preoccupied with explaining, chronicling, and commemorating its beginnings. It has developed a number of different narrative accounts about how, why, and where it began, some of which resemble what anthropologists term 'myths of origin.'(Fox and Swazey, 2008: 123)

Fox and Swazey identify a number of these 'origin stories'. One kind involves 'a technologically driven genesis'. That is, the 'moment of creation' is seen to be linked to advances in 'technologically based advances in biomedicine'. An example they cite is technological advances in life support—cardiopulmonary life support systems and organ transplants using cadaveric donors—which challenged established definitions of death and supported the 'use of irreversible coma as the criterion for determining whether and when a person had died'. Another is the long-term use of the dialysis machine for those with kidney failure, which raised questions about which patients should have access to this scarce and potentially life-saving treatment (Fox and Swazey, 2008: 23; see also Rothman, 1991: 149–153). Another kind of origin story focuses on issues which provide a catalyst for bioethics' emergence. These include constellations of particular issues such as technological developments and fertility control debates (Fox and Swazey, 2008: 24–25). The third kind of origin story pays cognisance to particular events, such as 'the medical experiments conducted on prisoners in Nazi concentration camps during World War II, the subsequent trial of Nazi physicians at Nuremberg, and the promulgation of the Nuremberg Code of human experimentation in 1947 by the tribunal's American judges' (Fox and Swazey, 2008: 25).

According to Fox and Swazey (2008: 27), there are also origin stories that highlight the 'organizational and linguistic moments of creation'. These may include the establishment of major institutions, such as the Hastings Centre in 1969, the Joseph and Rose Kennedy Institute for the Study of Human Reproduction and Bioethics in 1971, and the coining of the term 'bioethics' in 1970. Finally, there are more complex accounts, which point to 'a multiyear, multicausal gestation'. That is, contrary to the 'big bang' moment of creation story, bioethics is seen to emerge over time and as a result of an array of influences. For example, in one account, the years 1966 to 1976 were decisive because they began with Henry Beecher's exposé of abuses in human experimentation through to the case of Karen Ann Quinlan which is judged critical in 'the style and substance of medical decision making' (Rothman, 1991, citing Fox and Swazey, 2008: 28). Fox and Swazey favour the latter and in presenting their own origin story, identify a number of crucial 'phases' spanning a lengthy period, beginning in the 1950s, with the emergence of a number of medical research and health care issues, through to the early to mid 1970s, when various developments, including new bioethics journals and teaching programs, signalled that bioethics had 'arrived' (2008: 32–33).Bioethics histories tend to present a teleological view of history in that the present is portrayed as somehow an inevitable outcome of a series of antecedent events or circumstances. Historical narratives are used to explain the purpose of the field and to confirm the validity and rightness of current institutions and practices. Despite their different foci, the above histories (including, arguably, Fox and Swazey's) portray the development of bioethics as fundamentally progressive, with bioethics conceived as a product of increasingly enlightened thought, and representing a clean break with the past. According to Kim Little, in her study of the history of bioethics, 'clean break accounts dominate the historical self-image of bioethics' (2002: 49). In her view, such clean break accounts provide bioethics with 'a clear, conceptually simple relationship to the past', allowing proponents to highlight the novel character of bioethics, and its distinction from earlier forms of medical ethics (2002: 49). These novel aspects include the focus on increased respect for individual autonomy, the creation of new medical technologies, the involvement of actors outside academe (e.g. lawyers, policymakers), the increased focus on allocation of resources, and the secularization of medical ethics (2002: 51).

In this 'clean break' account of its evolution, bioethics is seen to involve the incremental accumulation of knowledge about the 'rights and wrongs' of policies, practices, and programs, and the gradual uncovering of truth through reason. Problems requiring bioethical reasoning and action arise as a result of the actions of altruistic individuals, or social, economic, and technological changes that predispose to a more progressive view on the institutionalised practices that become defined as problematic. Interventions are informed by the belief that problems can be rationally managed and controlled, absolutely, through the application of abstract reasoning.

Such histories tend to be descriptive—focusing on particular innovations, events, problems, programs, and personalities—rather than explaining how a particular set of conditions or circumstances predisposed to the emergence of bioethics as a field of knowledge and practice. They mostly lack a sense of the dynamic interplay of economic, political, and social factors in the evolution of knowledge and fail to offer insight into the role played by contending interests in the establishment of policy agenda.

Most histories of bioethics arguably serve to buttress the epistemic authority and legitimacy of bioethics, in much the same way as the histories of particular professions, such as medicine and law, serve to confirm their status and power. As such, they do not substantially advance our understanding of why a certain form of reasoning and associated expertise and domain of practice should emerge and become increasingly pervasive in the late twentieth and early twenty first centuries. Less still do they cast light on the conception of society and of the ideal social order that informs interventions. The main 'schools' of philosophical thought from which 'mainstream' bioethics purportedly draws its inspiration—utilitarianism, virtue ethics, deontological theories—are centrally concerned with the notion of the ideal or 'good' society, virtuous being and action, regard for the other, and so on. However, in their effort to gain legitimacy for the field of bioethics, scholars have developed abstract, universal principles upon which there can be broad agreement, and that will be viewed as 'non-partisan', thereby 'screening out' contextual issues and normative, potentially contentious, i.e. political, questions. Issues of power and politics and political economy that underlie inequalities and injustices are consequently largely 'framed out of the picture'. Bioethical issues very often become matters of technical-administrative determination and, as such, are subject, in the final analysis, to the constraints posed by economic and bureaucratic imperatives, limitations of time, vested interests, and entrenched ways of thinking and acting.

Conceptions of Technologies in Histories of Bioethics

As mentioned, histories of bioethics often focus on the significance of technological advances. Bioethics is seen as responding to various dilemmas and definitional quandaries posed by particular innovations, for example, definitions over the meaning of death arising from the use of technologies of life support and questions over access arising from the use of scarce, life-saving innovations such as dialysis machines. Concerns about 'tampering with nature', arising for example with cloning or the harvesting of embryos in stem cell research, and about infringing privacy or creating insidious new forms of surveillance arising from the use of new genetic technologies, are common. In most cases, 'technology' is conceived as outside 'society', or 'pre-social', and thus unaffected by politics and power. However, as sociologists of science and scholars of science and technology studies emphasise, technologies are *always already* social, in that they are produced and are ascribed meaning through social interaction and reflect social values and interests.

Science and technology studies' concept of 'co-production' highlights the fact that science and society stand in a mutually dependent relationship, with each profoundly shaping the other. Further, as histories of science and technology underline, all technologies embody assumptions about users and the uses to which technologies will be put (Hård and Jameson, 2005). Cars, the internet, mobile phones, and household appliances have all been designed with particular applications in mind and with a view to how individuals and particular groups will interact with the technologies. However, 'users' may view and use technologies in ways unimagined by their designers (Oudshoorn and Pinch, 2003). For example, the mobile phone and the internet have created the conditions for the emergence of new virtual communities, which interact in ways largely unforeseen by those who developed these technologies (see Chapter 5). 'Users' ascribe meaning to technologies in line with their own constructions of their social world (Schutz, 1972) and class-shaped dispositions (Bourdieu, 1986). It therefore needs to be acknowledged that 'users' and 'technologies' are always 'co-constructed' (Oudshoorn and Pinch, 2003).

The inescapably social dimension of technologies is captured by the term 'socio-technical', a term widely used in science and technology studies. 'Socio-technical' includes the representations of technologies in science, policy, and the broader culture. Although often overlooked in bioethics debates, the public representations and expectations of technological development—of the nature of the technologies, and when and how they will be used and by whom—are crucial in shaping the path of the development of technologies. Expectations exert a powerful influence on the development of science and technology *in general*. Much, if not most, scientific research would be difficult to sustain without a guiding vision of where it is heading and of the economic, health, and social benefits that will accrue. However, as I argue in the book, and especially in the next chapter, overly optimistic expectations carry dangers. The political and practical significance of expectations, I contend, needs to be more fully acknowledged, and they should be ascribed a more prominent place in discussions about the implications of the biosciences and biotechnologies.

CRITICAL PERSPECTIVES ON BIOETHICS

Over the last decade of so, a growing chorus of critics—social scientists, and particularly sociologists, feminist scholars, and disability scholars—have pointed to the deficits of 'mainstream' bioethics, especially the neglect of social contexts and the often unrecognised impacts of bioethics 'ways of knowing'. The contributions of scholars are diverse but very often arise from empirical research in particular areas of health and medical care, such as the pharmaceutical industry (Abraham, 1995; Hedgecoe, 2004b), genetics in healthcare (Petersen and Bunton, 2002; Bunton and Petersen, 2005), biobanks (Corrigan and Tutton, 2004; Gottweis and Petersen, 2008), participation in clinical drug trials (Corrigan, 2003), transplantation of body organs (Fox and Swazey, 1978, 1992), neo-natal care (Guilleman and Holmstrom, 1986), stem cell research (Salter and Salter, 2007; Williams and Wainwright, 2010), and the neurosciences (De Vries, 2007).

Sociological Perspectives

A significant area of critique arises from the sociology of bioethics, contributed to largely by health sociologists in the UK, Canada, and the US. Some scholars are based in medical schools and researching the ethical issues arising from clinical practice, or in sociology departments teaching and researching in the field of the sociology of health and illness or science and technology studies. In the UK, a considerable impetus to work in this field has been the Economic and Social Research Council's (ESRC) research programmes: the Innovative Health Technologies Programme, the Genomics and Society Programme, and the Stem Cell Programme. For a period in the late 1990s and early 2000s, the Wellcome Trust supported a more sociologically-oriented bioethics research and funded postdoctoral fellows and project grants through its Bioethics Research Programme. A number of doctoral and post-doctoral students whose work has been funded by one or more of the above programs have gone on to develop successful academic careers. Momentum for research in this field has been sustained through a number of social science research centres focusing on the biosciences that have been established in recent years: BIOS at London School of Economics; Centre for Biomedicine and Society (CBAS) at King's College, London; and Policy, Ethics, and Life Sciences (PEALS) at Newcastle Upon Tyne. Similarly, in the US, the Human Genome Project in the 1990s contributed a significant amount of funding (5–7 percent of the total budget) to social science, including sociological research *in* and *of* bioethics, as well as the establishment of a number of courses exploring pertinent issues.

Sociologists have highlighted the assumptions underlying bioethical reasoning, drawing attention to this field's 'blind spots', and often proposing ways forward for a 'better' or 'more enlightened' kind of bioethical practice. De Vries, et al. (2007) have noted what they referred to as the 'trained incapacity' of bioethics scholars to recognise the significance of the socio-cultural context. De Vries, et al.'s book includes a number of chapters that argue the need for a more sociologically-informed, empirically-grounded approach within bioethics. Either explicitly or implicitly, among these contributors and other recent writers, sociology is ascribed a 'saviour' role, offering a fuller or 'more complete picture' of the issues that bioethics systematically neglects (see, e.g. Corrigan, 2003; De Vries, 2003; Hedgecoe, 2004a; López, 2004). Bioethics is perceived to be overly reliant on moral

and analytic philosophy which has, in the words of López, 'created a selec-
tivity towards a formalistic, procedural, disembodied and universalistic
way of identifying and resolving bioethical dilemmas' (2004: 878).

In the view of López, bioethics should be of interest to sociologists
because of 'its' alignment with core social values. He asks whether sociology has a role in 'saving' bioethics and believes that ethnography can 'add
something to bioethics' by empirically grounding its abstract concepts in
the contexts of interaction where the rules of ethical behaviour are embedded. Similarly, in her interview-based study of participants in clinical drug
trials, Corrigan criticizes the approach to informed consent which is 'premised largely on the autonomous individual and his or her rights, with little
or no conception of the social aspects' (2003: 770). As Corrigan notes, this
'empty ethics' model (her term) de-contextualises the process of consent
which is reduced to a rational choice model of action which presupposes
that decision-making is simply a matter of adequate provision of information and time (2003: 770). Corrigan's sociological analysis presents a more
complex view on the implementation of informed consent than is portrayed
in bioethics guidelines and policies. The picture, she argues, is complicated
by patients' expectations of research, the nature and severity of their illness, and the power relationship that always exists between the patient and
the clinician that shapes interactions.

According to these sociologists, the 'incompleteness' of bioethics can
and should be addressed by greater reference to or the incorporation of the
social sciences, and particularly sociology. In this view, bioethics needs to
overcome its 'empirical deficit' by grounding concepts in empirical research
employing, in particular, ethnographic methods, and bioethicists need to
develop and exercise their 'sociological imagination'. In arguing for a 'critical bioethics', for example, Hedgecoe proposes that 'The problems, dilemmas and controversies analysed come from looking at a particular setting
(e.g. the clinic), talking to participants and taking note of what they say'
(2004a: 135–136). In promoting this 'bottom up' approach to ethics, he
sees social science research as providing the starting point for bioethical
enquiries. However, as he explains: 'This does not mean that philosophers
have to become social scientists; simply that if they are interested in the
ethics of a particular technology, their first port of call should be the social
science literature about that technology, rather than the standard bioethics debates' (Hedgecoe, 2004a: 136). While such research may support the
principalist approach, it may equally challenge analysts' established theoretical frameworks and lead to their modification or, in some cases, their
abandonment (2004a: 137).

The call to employ ethnography and other social science methods has
indeed been taken up among some bioethicists to explore how health professionals and patients and families experience and respond to bioethical
issues (Fox and Swazey, 2008: 179). A study published in 2006 by Borry, et
al. (2006), involving a survey of nine peer-reviewed journals in bioethics,

found that there had been an increase in the number of multiple-author articles and a decrease in the number of single-author articles published between 1990 and 2003, which they attribute to the growth in the amount of bioethics research involving an empirical design. As Borry, et al. observe, 'Articles reporting empirical research are 10 times more likely to present more than one author in the byline that others' (2006: 218). However, as Fox and Swazey observe, bioethicists have tended to adopt a rather facile conception of ethnography, failing to recognise the range of qualitative methods that are involved and have overly high expectations about what the approach can deliver in terms of transforming bioethics and bioethicists (2008: 180–182). As they note, 'Ethnography *per se* does not have these transfiguring capacities', and 'if bioethicists who are not trained ethnographers engage in this sort of research, and carry such misunderstandings into the field with them, they might do more harm than good to the prospects of advancing social and cultural thinking in bioethics' (2008: 182).

Feminist Perspectives

Feminists share similar concerns to sociologists about bioethics' neglect of the social context, but their criticisms tend to be more specific and have a stronger normative basis. Their concern is not so much the lack of a sociological imagination and an 'empirical deficit' but rather lack of reflection on gender biases in its frameworks and inattention to issues of global justice and its underpinnings. Feminist critiques of bioethics are diverse, which is hardly surprising in light of the multiplicity of *feminisms*, including liberal, cultural, radical, marxist, and postmodern/poststructural perspectives. However, they share a number of broad criticisms of 'mainstream' bioethics' 'ways of knowing'. Feminists are critical of the overly narrow focus of bioethics, its heavy reliance on principlism, and the marginalisation of questions of concern to feminists, especially questions of justice. The knowledge of bioethics is seen to suffer from an 'androcentric' bias, especially in relation to conceptions of the body. For example, women are often excluded from clinical drug trials, even for non-gender-specific diseases, often on the basis that their hormones can present a 'complication'; namely, that women's hormonal patterns can make it difficult to discern the effects of a drug or procedure being studied (Little, 1996: 4). As Little notes, this reflects 'confidence in the neutrality of treating the male body as the norm' (1996: 4) and the view that women's hormonal pattern distorts evidence concerning the true effect of a drug or procedure and hence is something to be ignored rather than regarding it as an important factor in its own right.

Feminist scholars have taken particular issue with bioethics' focus on autonomy and individual rights. The standard conception of autonomy, it is argued, has been developed within the context of patient–practitioner and researcher–subject relations which is both limited in its sphere of application and pays insufficient attention to the contextual factors shaping interactions. It has, for example, obscured the importance of the *groups* to which individuals belong, including those based on gender (Wolf, 1996: 17). It is not that 'mainstream' bioethics has been completely blind to the moral significance of groups—differences of age, group inequalities in access to healthcare provision, and so on have been subjects of concern to bioethicists—but the *priority* focus has been on 'the patient' or 'the research subject' which has obscured differences (especially of gender) among patients and subjects (Wolf, 1996: 18). As many feminists emphasise, decision-making always occurs within *relationships* and so is never fully autonomous in the ways portrayed in the articulations of bioethical principles. Bioethics' focus on autonomous, rational decision-making reflects the 'androcentric' reasoning that is deeply inscribed in philosophy, which separates reason from emotion, associated with men and women, respectively (Little, 1996: 6). A model of male rationality is placed at the centre of analysis, which denies the realities of women's lives, particularly their relations in the private sphere (e.g. with sexual partners, between parents and children, carers and the cared for) which are unequal, invisible, and cannot be governed by codified rules (Little, 1996: 7). A feminist bioethics eschews notions of rationality and human rights articulated in the name of 'man' in favour of an analysis of women's different, 'situated' experiences and perspectives and gives prominence and understanding to practices and behaviours that are harmful, exclusionary, and discriminatory (see, e.g. Wolf, 1996: 22–26).

The above major lacuna in 'mainstream' bioethics, many feminists argue, calls for nothing short of the field's radical transformation. Many articles written by feminists on bioethics seek to reorient bioethics by shifting attention to issues of justice and agenda-setting rather than the current *reactive* approach to the field which skews responses in line with agendas established by others (e.g. Donchin, 2001; Little, 1996; Rawlinson, 2008; Sherwin, 2008; Wolf, 1996). Sherwin describes the limitations with the reactive approach of 'mainstream' bioethics:

> Much of the work of bioethicists involves reacting and responding to problems or positions that others pose for us through personal, institutional, or media requests. The difficulty with this pattern, of course, is that when we take up challenges initiated by others, we use time and energy that might be used to pursue other important ethical questions. The opportunity costs associated with allowing our schedules to be filled by questions set by others are significant as we can see when we consider the stakes of the global problems that we are neglecting. . . . (2008: 19)

Sherwin points to the dangers of entering discussions of topics initiated by others: one is likely to begin with and get caught up in the framework initially used to pose the problem (2008: 19). She uses the example of reproductive ethics where debate has focused on the moral status of human foetuses or embryos (see Chapter 4). Focusing on this as the primary and most important ethical question, she argues, has diverted attention from issues that 'most feminists believe should be at the forefront of public debates and policies' (Sherwin, 2008: 20). By focusing on the question of the moral status of the embryo, liberals and moderates can play into the hands of conservatives who insist that this is the key issue for deliberation. It can lead to the neglect of the rights of the pregnant woman and the prioritising of the rights of the foetus over the rights of the woman (Sherwin, 2008: 20–21).

Growing feminist dissatisfaction with 'mainstream' bioethics over the last fifteen years or more has led to a number of feminist-led initiatives, including the establishment of The International Network on Feminist Approaches to Bioethics (FAB) (in 1992), involving a global network spanning twenty-six countries on six continents, biennial FAB-sponsored conferences and resulting books focusing on feminist perspectives on bioethics, and the launch of a new journal, *International Journal of Feminist Approaches to Bioethics*, in 2008 (Donchin, 2008). It is clear from statements in this journal, pertaining to the advancement of global justice, 'taking into account the specificity and irreducibility of the experiences and bodies of women' (Rawlinson, 2008: 3), that the aim is to go beyond 'saving' bioethics to radically reconceptualising the field. Feminist bioethicists not only reject the use of abstract universal norms and universal moral principles. In focusing on human rights and promoting the vision to create a 'nonhierarchical human community committed to mutual support and optimization of the health and wellbeing of all' (Donchin, 2008: 156), they are outlining a much clearer conception of the 'good' society than is evident in either 'mainstream' bioethics or sociological critiques of this field. In particular, feminist scholars have promoted the concept of 'global bioethics', one that does not erase differences, but rather (in the words of one writer) 'takes into account the diversity of peoples and cultures in our world, as well as the fact that our planet and its resources provide the bases for all our lives' (Tong, 2001: 27). This is an explicitly political agenda, which acknowledges that it is impossible to 'sit on the fence' or refuse to make judgements about developing nations, systems, and practices that would be viewed as morally wrong or oppressive in their own countries (Tong, 2001: 28). Such a position, Tong argues, is antithetical to feminist politics and action and to furthering *just* global policies; namely, 'international policies that distribute freedom and wellbeing (in the form of goods and services) equally among all the individuals they affect' (2001: 28).

Disability Studies Perspectives

As with feminist scholars, disability scholars do not hold a unitary perspective on bioethics. However, they tend to share a number of concerns about this field, particularly the dominance of principlism which is seen to narrow the focus of debate and action and to operate in an exclusionary way. A common view is that bioethics principles derived from moral theory are far removed from the lives of disabled people who are directly or indirectly affected by medical decision-making involving bio-technologies. Disability scholars' and activists' particular concerns arise from the context of clinical decision-making, especially pertaining to the use of prenatal genetic screening for potentially disabling conditions, and assumptions about the quality of the lives and the rights to life of those who are severely disabled. As Scully argues in her recent book, *Disability Bioethics*, 'Bioethics' involvement with disability has predominantly concerned moral judgements about the quality of life' (2008: 52). As she notes, 'These judgements assess what it is like to live with an impairment and decide whether, according to that assessment, further steps are morally permissible or not' (2008: 52). By imaginatively 'putting one's self in someone else's shoes', one is able to understand the experiences and perspectives of the disabled person. However, this proves difficult in practice, as shown by 'empirically observed discrepancies between reported and projected accounts of what it is like to live with an impairment' (Scully, 2008: 56). Social positioning is significant in experiences of impairment and disability and in understanding life events and choices (Scully, 2008: 56).

The growing use of genetic testing in healthcare (see Chapter 5) has heightened fears about discrimination against people with disabilities which may be reinforced by actions based upon the prescriptive applied approaches of bioethics. According to some writers, screening for genetic-based disability conditions (e.g. Down syndrome) is inherently discriminatory because it implies that people with disabilities lead unworthy and poor quality lives. Disability scholars and activists worry about the practice of offering prenatal tests for genetic markers of impairment and the subsequent selective offer of abortion, which is supported by the biomedical community. Concerns have been expressed about the eugenic implications of the routine exercise of 'choice' in prenatal decisions, namely, the decision to terminate in cases where a genetic defect is found through genetic testing (e.g. Bailey, 1996; Shakespeare, 1998). In arguing for patient autonomy or 'pro-choice', bioethicists have been seen as supporting the rights of the mother or parents (to abort) against the rights of the unborn, potentially disabled child. Utilitarian bioethicists in particular have been criticised for their assumptions about the value and quality of the lives of disabled children. The potential existence of infants born with intellectual disabilities, for example, is seen as less valuable than the existence of 'normal' infants (Vehmas, 1999). The utilitarians, Peter Singer and Helga Kuhse, for instance, have been taken to task by disability scholars for their argument that in certain circumstances infants with severe disabilities should not be allowed to live (Newell, 2006: 272).

Bioethics debates, especially those arising from utilitarianism, are seen as mostly *uninformed* by the experiences of disabled people themselves and their views on the meaning and quality of their lives. As Newell (2006) argues, despite accounts of severely disabled people themselves of living full, happy lives, bioethicists have largely failed to repudiate the view espoused by Singer and Kuhse and supportive utilitarians. Discourses of tragedy and catastrophe are seen to pervade bioethicists' understandings of disability (Clapton, 2003). As Goering argues,

> Bioethics debates are often disturbing to disabled people, and to disability rights activists in particular, given the way they presume that living with an impairment is inevitably bad, or much worse than living a 'normal' life. They tend to overestimate the degree of difficulty faced by people with impairments and incorrectly identify the main causes of the difficulties. (2008: 125)

As with feminists, disability scholars see autonomy as being much more complex than portrayed in the individualist model of bioethics discourse and as being always *relational*. As Ho observes, the focus on individual competence and capacity in decision-making denies 'the impact of the larger social structure and ideology in determining potential patients' value framework and available options' (2008: 197). Bioethics' conceptions of autonomy are based on the didactic relationship between two individuals, one of whom is dominant (e.g. the physician) and the other subordinate (e.g. the patient), and fail to acknowledge that choices do not occur in isolation. Drawing on the insights of feminist writers such as Nancy Hardstock and Iris Marion Young, Ho argues that patients' decisions are never solely individual decisions; they 'are embedded within a complex set of social relations, practices, and policies that structure an individual's selfhood and can significantly affect people's ability to exercise autonomy with respect to their choices' (2008: 195).

RETHINKING BIOETHICS

Taken as a whole, the above critiques emphasise the significant limitations and implications of the approaches offered by mainstream bioethics. In particular, they highlight the generally narrow focus of bioethics' 'ways of knowing' due to its heavy reliance on certain disciplines and perspectives, particularly principle-based moral philosophy. Bioethics is a product of a particular time (1960s and 1970s) and place, namely the US, and reveals the worldview and interests of relatively powerful, Western elite groups. It has evolved primarily from clinical and research contexts, and thus focuses on issues relevant to those contexts, for example, the clinician–patient relationship and the researcher–subject relationship, patient autonomy, confidentiality, informed choice, and so on. However, increasingly, bioeth-ics' knowledge has been applied beyond those contexts to address issues and dilemmas posed by new biotechnologies which have potentially far reaching, long term impacts on society. It is being applied *globally* to issues or problems confronting peoples that did not previously call for delibera-tion via bioethical frameworks and expertise.

As critics point out, bioethics concepts and principles reflect a West-ern liberal view of the world that, in its application, arguably serves to legitimise rather than challenge dominant relations of power. It reflects a historically and culturally specific concept of the human subject—as a rational, independent decision-maker—and of human freedom—conceived as absence of constraint and ability to pursue one's own interests. Bioeth-ics' conception of citizenship is one that became increasingly dominant from the 1970s, namely, that of neo-liberalism. In this conception, 'soci-ety' is either downplayed or ignored in favour of the assumed interests of the 'individual'. In the effort to develop abstract universal principles that apply without favour to individuals in similar situations, the notion of the broader public good has been sidelined, and the views and experiences of particular groups (women, people with disabilities, minority ethnic groups, religious minorities) have been excluded. Bioethics' inattention to socio-cultural contexts and to empirical evidence, noted by critics, is reflective of these views on self and society.

Cross-cultural analyses reveal how local histories, cultures, values, politics, and legislation shape conceptions of 'ethics' and ethical practices. For example, the Filipino conception of bioethics is wider in scope than that which prevails in Europe and the US, referring not to the applica-tions of abstract principles to certain situations, but rather as 'a form of life embedded in its socio-cultural context that covers all aspects of life' (Kae-lin, 2009: 43). Further, ethical practices may depart from articulated ethi-cal principles for a variety of culturally specific reasons. In China, while the guidelines and regulations governing the ethical aspects of biomedical research are not essentially different from those in Europe or the US, they are not enforceable by law and are often not implemented. In that country, the publics' historical mistrust of politicians and scientists adversely influ-ences views on the donation of tissue samples and blood to scientists. Low levels of education among many people, especially in rural areas, limit pub-lic debate on the social impacts of biomedicine and undermine the notion of informed consent as applied to research participation. Further, endemic corruption, for example among biotechnology companies works against basic patient rights. Consequently, despite increasing efforts in recent years to regulate activities in the biomedical sector, practice in research and medical institutions is not in accordance with ethical principles and exist-ing bioethical guidelines (Hennig, 2006: 850–851). Notwithstanding an apparent willingness among many Chinese researchers in the life sciences to accept more restrictive regulations in exchange for more effective ethical guidance, legal and administrative systems are unprepared to deal with issues in a comprehensive way (Döring, 2003: 44). Evidence from other cultures, for example, Japan and other parts of East Asia, underlines the variability of ethical conceptions and practices. Cultures may differ according to their conceptions of autonomy and justice and perceptions of science, risk and trust, and consequently the ethical dilemmas that arise (e.g. Asai, et al., 1997; Kaelin, 2009; Suda, 2008).

Radically new approaches and mechanisms for advancing broad social justice objectives are needed for assessing bioscience projects and biotechnology and clinically-based innovations that have broad *population* level applications and impacts, that involve *control over life processes* (e.g. reproduction and regeneration), and that call for *longer-term* investments and the participation of publics, which is the case with most contemporary biotechnology research. Given the potentially profound impacts of biotechnologies presently utilised, in development, or 'on the horizon' on conceptions of the body, self, and society, there is a need for wide deliberation on the best means for appraising innovations either very early in the technology development phase or ideally *before* technologies begin to be developed. This calls for nothing short of a fundamental change in the current predominant science–society relationship, and in particular in the mechanisms of democratic participation. What is needed is greater deliberation on the desired ideals of citizenship with specification of the rights and responsibilities of different constituencies ('stakeholders') in and between societies and acknowledgement of the different interests at stake in science and technology developments (see Chapter 7). The incapacity of bioethics to respond critically and usefully to the challenges posed by the biosciences and biotechnologies has become increasingly evident as developments in a number of fields gain momentum.

OUTLINE OF THE REMAINING CHAPTERS

Chapter 2 examines the important role played by expectations in the process of technological innovation, making reference to developments in the fields of genetics and personalised medicine and stem cell research, respectively. The political significance of expectations has been largely ignored by bioethicists and other observers of biotechnology developments, who tend to focus on the problems posed by technologies *as they arise*. Indeed, bioethics tends to unquestionably accept these expectations. Given that expectations are socially produced and sustained, being supported by various groups that have a vested interest in the fulfilment of particular technologically-enabled futures, their role and implications, I argue, deserve much closer attention.

Chapter 3 focuses on recent biotechnology innovations of growing international significance, namely, biobanks. I examine the promises of biobanks and the challenges they pose to bioethics. In recent years, many countries have begun to develop biobanks in the expectation that they will deliver substantial economic and health benefits in the future. Given their perceived role in the emergent bioeconomy and in pinpointing the genetic contributions to disease, they have achieved considerable support from governments and science groups. However, like the other innovations considered in this book, they entail many uncertainties. In this context, bioethics has served a legitimising role, helping to engender public consent for what are in effect large infrastructure projects of long duration. In the chapter, I explore the role played by bioethics in governing biobanks and the limitations of its concepts, and particularly 'consent', in addressing the challenges that biobanks present.

Chapter 4 turns attention to the highly contested field of stem cell research. Despite the broad range of normative and justice issues raised by this field, bioethics debates have focused narrowly on a limited array of issues that have been defined largely by scientists and their critics and portrayed in the media. Attention has focused on the moral status of the embryo and the rights of donors. In their preoccupation with the technologies themselves and with applying abstract principles to discrete issues, bioethicists have failed to engage with broader justice questions. Despite the many uncertainties of the field and the limited applications to date, the research effort has not slowed, being heavily supported by governments that are keen to reap the expected future benefits. It is in this context that bioethics plays a mediating role in an effort to reconcile often irreconcilable values and positions. The chapter examines some recent sociological contributions to this field and some fundamental unaddressed problems.

Chapter 5 investigates the challenges posed by genetic testing, especially that advertised via the internet. The recent rapid growth of 'direct-to-consumer' advertising of genetic tests and other medical tests and treatments is a manifestation of the profound changes under way in health and healthcare at the international level. This is an area where technological developments are running far ahead of ethical and regulatory responses. Discussions about the implications of genetic testing thus far has been restricted by a focus on a limited array of issues, such as how best to achieve 'informed consent' and how to ensure 'non-directive' genetic counselling in the case of those who have been tested. This has served to obscure the politico-economic and socio-cultural implications of the increasing influence of genetic information in healthcare and other areas of life. The chapter draws attention to the limitations and implications of the concepts of autonomy and informed consent within the field of genetic testing and genetic counselling, drawing on the work of feminists and other critical scholars.

Chapter 6 examines the role played by bioethics in the governance of nanotechnologies. Nanotechnologies are the subject of very high expectations, with applications predicted in many areas, including health and medicine. However, nanotechnologies involve considerable uncertainties, particularly because applications rely on their convergence with other technologies. Uncertainty surrounds the novelty of the field, the specific applications, the biophysical risks, and, most importantly from a governance perspective, the social responses. Because the area is emergent and the applications diverse, the 'user' groups are various and cannot be easily specified before applications are settled. Thus far these technologies have not been the focus for citizen organisation and activism as have genetic therapies and stem cell treatments. Achieving consent for technologies that have yet to emerge and that involve so many uncertainties presents acute problems for governance. In this context, I argue, ethics and specifically bioethics, along with the social sciences, have been called upon to help engender consent and legitimacy for the field. In particular, the knowledge from these disciplines has assisted in translating the uncertainties of nanotechnologies into the familiar language of risk, thus making them *governable*.

Finally, in Chapter 7, I conclude the discussion by offering some suggestions for how to move 'beyond bioethics'. Bioethics is a product of a particular culture (US) and time (1960s and 1970s) and developments in the biosciences and biotechnologies since then, and especially since the beginning of the twenty-first century, call for new tools of analysis and critique. Despite substantial criticisms of the field, particularly from sociology, feminism, and disability studies, and its increasingly obvious limitations in practice, bioethics seems resilient. This is because bioethics is closely aligned with dominant interests and has achieved considerable legitimacy and support within government, business, and other spheres. In the chapters, I highlighted some unacknowledged implications of how bioethics is deployed in practice which, I hope, assists in its demystification and the lessening of its power. However, moving beyond bioethics calls for more than critique and deconstruction. It entails developing mechanisms for democratising science and technology, posing new questions, developing new concepts, and utilising the insights from a range of fields.

2 Bioethics and the Politics of Expectations

> It's June 2018. Sally picks up a handheld device and holds it to her finger: with a tiny pinprick, it draws off a fraction of a droplet of blood, makes 2,000 different measurements and sends the data wirelessly to a distant computer for analysis. A few minutes later, Sally gets the results via e-mail, and a copy goes to her physician. All of Sally's organs are fine, and her physician advises her to do another home medical check up in six months. This is what the not-so-distant future of medicine will look like. Over the next two decades, medicine will change from its current reactive mode, in which doctors wait for people to get sick, to a mode that is far more preventive and rational.
>
> (Hood, 2009: 50)

Predictions about the future benefits of medicine are rife in the field of biotechnology. The expressed confidence in this scenario—that medicine *will* change in the direction outlined—is common in both recent scientific and popular literature. In particular, the expectation that biotechnology will help make medicine more predictive and personalised in the future has been a key theme in both academic literature and policy documents and has arguably provided a major spur to innovations and influence on health policy decisions. It is also expected that in the future medicine will be more preventive and participatory; hence Hood's designation 'P4 medicine'—predictive, personalised, preventive, and participatory (2009: 50). Many emerging technologies are assumed to enhance individual choice and assist in developing interventions that will be oriented to the prevention of illness. Breakthroughs in human genomics, combined with the convergence of genetic technologies with other technologies, including nanotechnologies and digital technologies, scientists claim, will permit the systematic pre-detection of disease—a scenario depicted above. Despite acknowledged uncertainties about how and to what extent technologies will converge and what opportunities (and dangers) this will present, proponents of new technologies express little doubt that 'the public' will in time derive benefits of the kind described.

Expectations of future public benefit have been a considerable 'driver' of technological innovation in general in many contemporary societies—in forging connections between different communities of interest, in eliciting funding for research, and in mobilising action, and thus in shaping the future. However, the social role and implications of expectations have been largely ignored by bioethicists and by those who draw on bioethics' ideas in assessing or responding to the impacts of biotechnologies. In the bioethics literature, few questions have been asked about where these expectations originate, how they are sustained, and who they benefit. This blindness to the socio-political significance of expectations, I suggest, reflects a broader bias in bioethics' 'ways of knowing', in particular the concern with developing abstract principles and the neglect of social contexts and processes. As a number of critics of bioethics have noted (see Chapter 1), this field has been preoccupied with *problems posed by technologies as they arise.* Bioethicists have debated the implications of particular innovations (e.g. IVF, cloning) exploring such questions as whether bioethics should adopt a 'pro-active' approach in addressing the challenges posed by technologies before they come into widespread use or offer its assessments of the benefits and implications once they have developed (see Brody, 2009). There has been little concern with the values and interests that shape science and technology itself and with the ways in which science and technology in turn shape identities, relationships, networks, and values. Claims about the benefits and impacts of technologies are rarely based on empirical research; for example, pertaining to how different groups define and use (or have in the past used) technologies and are affected by them in their everyday lives.

As Paul Farmer and Nicole Campos note, despite the claims to objectivity, science and medicine are social in nature: 'The research problems and questions we choose to pursue, the methods we follow, and the practical applications that result from scientific discovery are all determined by society's priorities and interests' (Farmer and Campos, 2004: 3).The tendency of bioethics to focus on 'technology' in the abstract has led to the neglect of broader issues of social justice (Brody, 2009; Farmer, 2003; Farmer and Campos, 2004). Questions about the inequitable social impacts of priorities in research funding, which favour technological solutions over socio-political solutions, are rarely asked. As Brody (2009), and Farmer and Campos (2004) observe, bioethics' preoccupation with technologies reflects the priorities of rich developed nations, where technological innovation is highly valued and mostly unquestioned as the basis for economic and social wellbeing. For much of the world's population, more pressing issues, like gaining access to clean water and adequate housing and creating a sustainable economy, call for urgent attention. When examined in a global context, the whole field of bioethics is seen to give *disproportionate* attention to emerging biomedical technologies (Brody, 2009: 205).

This chapter focuses on the politics of expectations—their role in guiding biotechnology innovations and their significance in shaping social priorities and social arrangements—a recurrent theme in the subsequent chapters of this book. It examines how bioethics' focus on the problems posed by 'technologies' as they arise, and on their resolution through reference to abstract principles and use of technical administrative mechanisms, serves to depoliticise issues. In recent 'ethical' deliberations on emergent biotechnologies, there has been little questioning of the social, economic, and political underpinnings of technological innovations and of the assumptions that guide expectations about their development and use. These are strong claims that will no doubt be contested by bioethicists and those who utilise bioethics' frameworks and principles with the best of intentions to help resolve the challenges posed by biotechnologies. However, this argument finds support in the work of a growing body of evidence, including that generated by sociologists, scholars in the field of science and technology studies, feminists, and disability scholars, a number of whom were referred to in Chapter 1. As I argue, expectations are sustained through diverse practices, including an array of activities covered by the term 'public engagement', a range of promotional activities undertaken by scientists and science proponents, scientists' strategic use of diverse media to convey a positive view of science, and scientists' and policymakers' recourse to bioethics knowledge and expertise to address issues that raise substantive questions. These practices, I contend, serve ideologically to help generate belief that new technologies will be of broad public benefit and that any risks that may arise can be adequately regulated, while obscuring the vested interests involved in developing and promoting innovations and the substantive questions concerning their ownership, access, and impacts.

Increasingly, 'public relations' has become a key tool in science for the management of expectations—in raising the profile of research and its applications, in 'hyping' the significance of research, and in 'educating' publics about where science is heading and the purported benefits. Bioethics has been called upon to assist in this endeavour, thereby performing a legitimatising role for science. However, technology proponents' use of public relations expertise is but the more visible aspect of the promotion of expectations, which also occurs through less public—and hence less visible—means, such the mobilisation of professional networks, the lobbying of key decision makers, and the nurturing of science–industry links. Clearly, the generation and sustaining of expectations is an inherently *political* process, in that it involves powerful actors drawn from science, business, and public authorities deliberating on matters oriented to affecting certain desired outcomes, especially those that are seen to produce economic benefit. These actors tend to share similar views on the value and significance of science, technology, and the economy. In this chapter, I examine some of the key mechanisms by which expectations are generated and sustained, and I discuss some of the attendant largely overlooked social, economic, political, and justice implications. In so doing, I focus on areas of technological innovation that have received a considerably high profile in science, policy, and the media in recent years; namely, genetics and personalised medicine, and stem cell technologies.

THE POLITICO-ECONOMIC DYNAMICS OF EXPECTATIONS

As noted, expectations play a significant mobilising role in the process of technological innovation in many contemporary societies. A strong vision of how technologies will develop and of the benefits that will accrue are necessary if proponents are to achieve the support of publics, funders, and policymakers in order to undertake research and develop technologies. In the case of medicines and therapies, the long time periods required for their development, involving human trials and the process of approval, can mean that many years may elapse before innovations become available in the clinic. Numerous factors may work against the realisation of technologies, including the failure to attract or sustain funding, shifts in policy priorities, changes in key actor networks, conflicts among influential individuals (e.g. scientists) or groups, unrecognised or unacknowledged flaws in the science that underpins research, and the resistance of publics. Given these potential (and frequent) barriers to the realisation of innovations, expectations can be difficult to sustain over the longer term. They therefore need to be supported through various reiterative practices, including the management of public representations.

Sustaining sources of funding is a key challenge facing proponents of new technologies. Policymakers and private investors need to be convinced of the benefits of particular technologies. Publics may have little interest in or understanding of science in general, let alone appreciate the potential of the envisaged technology. Any innovation will require a considerable investment of financial resources: in the basic 'bench-based' research, in the design and manufacture of the technology, in the trialling of the prototype or drug, and in the promotion of the product. In a market of competing demands for investment, research involving innovations that are seen to have the strongest prospect for generating healthy profits are most likely to attract investor funding. Consequently, the practical applications of 'breakthroughs'—the new therapies and treatments with potentially wide application—tend to be emphasised. However, the sources of funding can be tenuous, especially where there is a heavy reliance on the private sector, because shifting 'market sentiment' affects both short term and longer term private investment decisions—sometimes dramatically, as highlighted by the financial crisis of 2008–2009. The biotechnology sector, perhaps more than other sectors of the economy, is highly vulnerable to the rise and fall of the stock market due to the lead times involved in the development of new drugs. As I note below, an area of particular concern to the biotechnology sector—especially companies that are developing 'personalised' drugs—has been the impending expiration of many patents, which will enable generic drugs to compete with established brands. In the absence of new 'blockbusters', the biotech sector is finding it increasingly difficult to attract investors.

The exploitation and mobilisation of 'human capital' is a necessary condition for the promotion of expectations. Many actors and networks, often spanning the globe, play an active role in keeping visions alive and visible to a broad section of the population. Specifically, there is a need to enrol the support of key policymakers and funders, establish mechanisms for facilitating the sharing of research findings and developing the necessary research capacity, and build the institutional structures and regulatory environments required to move the research agenda forward in order to realise the vision. In some cases, technology innovations depend crucially on the 'harmonisation' of research efforts: to standardise methodologies; to develop protocols, procedures, and regulations that apply internationally; and to harness the required expertise. Increasingly, the 'big science' model operates in many fields of technology development, especially genetics, illustrated by the Human Genome Project (HGP), which involved the collaborations of many research teams spanning a number of nations working together to 'map' the human genome. In many cases, these collaborations involve the financial support of corporate interests, including pharmaceutical companies and other enterprises (e.g. the food and tobacco industries), that foresee benefits from genetic technologies. According to the UK's GeneWatch, the HGP had its genesis in collaborations between the tobacco giant, British American Tobacco (BAT), and the German pharmaceutical and chemical company, Bayer, in the joint funding of a research unit at Newcastle University 'which published numerous spurious results linking genes to lung cancer in a journal edited by its Director, Jeffrey Idle' (GeneWatch UK, 2010a: 1–2). The idea of developing genetic tests to predict which smokers would develop lung cancer (i.e. the 'genetically predisposed') suited the tobacco industry, GeneWatch UK claims, because it meant that 'smoking cessation efforts could be targeted at them so the rest of the population could continue to smoke' (GeneWatch, 2010a: 2; see also Doward, 2004).

In the so-called 'post-genomic' era, research endeavours such as biobanks also call for international collaboration to share ideas and experiences, research data, and ethical and regulatory protocols. The Public Population Project in Genomics (P3G) initiative is a visible expression of this, providing a repository of data on a range of matters of common concern, such as scientific rationale and methodologies, processes of consent, and ethics, governance, and public engagement issues relating to population-based biobanking (see: http://www.p3gobservatory.org/) (Accessed 12 August 2009) (see Chapter 3). Collaboration is often fuelled by competition, where groups of scientists, often drawn from a number of disciplines, link up in order to gain advantage over other groups in the 'race' to make the next big 'breakthrough', such as the development of a new therapy or 'blockbuster' drug. Considerable expectations attach to such big science projects, especially genetics research, stem cell research, and nanomedicine. These large research programs are more likely than smaller research programs to have the necessary financial resources, expertise (including public relations), and political links to sustain expectations over a long period of time.

Expectations are supported by various constituencies that have a strong stake in the fate of the biotech field: by the biotechnology and pharmaceutical industries that need the next 'big breakthrough' that will generate sufficient profits to satisfy their shareholders; by scientists whose careers and reputations depend on undertaking research seen as having public value; by universities and other research institutions that employ these scientists which need to establish the confidence of policymakers and funders in order to achieve and maintain their status as research leaders; by governments that expect that research will lead to the technologies that will contribute to the effort to contain burgeoning healthcare costs, particularly those associated with an aging population; and by individual patients and patient groups who hope that innovations will be developed that will relieve suffering. Rose and Novas (2005) refer to 'the political economy of hope' to emphasise the economic dimension of patients' hopes and expectations of 'saviour science' (see Chapter 5). I suggest that the political economy of hope is sustained through the activities of an extensive network of actors, all of whom have some degree of interest in sustaining the visions that guide research and development in the field of biotechnology. This includes the burgeoning number of social scientists, philosophers, bioethicists, and legal scholars whose careers and reputations have come to depend heavily on the fate of the programs of research that are oriented to assessing the ethical, legal, and social implications (ELSI) of innovations and/or developing strategies (e.g. 'public engagement') for managing the science–society relationship (see Chapter 6). Many, if not most, of the significant number of research centres focusing on the ethical and social implications of biotechnologies established in the US, UK, and Europe in recent years would cease to exist in the absence of strong expectations that promised new innovations will emerge in the not-too-distant future and give rise to novel social, ethical, and regulatory questions in need of their expert resolution. In short, a broad range of groups have a stake in the economy of expectations, with a number actively seeking to heighten expectations of particular technologies and their attendant benefits. The operation of some of the mechanisms that generate and sustain these expectations can be clearly seen in the field of genetics and personalised medicine.

GENETICS AND PERSONALISED MEDICINE

The excerpt at the beginning of this chapter represents a common vision of personalised medicine. The idea that one may be able to develop drugs that are 'tailored' to the genetic profile of the individual has considerable appeal among scientists, clinicians, and policymakers. This is because individuals are known to react differently to drugs. The phenomenon of adverse drug reactions (ADRs) has been identified as a major problem in medicine, with suggestions that between 5 percent and 30 percent of all patients receiving medical therapies have an adverse event of some kind (see http://www.virtualmedicalcentre.com/Treatments.asp?sid=65#C1) (Accessed 14 August

2009). According to one estimate, approximately half of drug-related injury arises through ADRs (Edwards and Aronson, 2000: 1259), a proportion of which results in deaths. For example, in Sweden, a recent study suggested that ADRs accounted for around 3 percent of all fatalities—the seventh most common cause (*Nature News*) (see http://www.nature.com/news/2008/080317/full/news.2008.676.html#cor1) (Accessed 14 August 2009). The economic and personal health costs associated with ADRs are considerable. A recent estimate indicates that in Britain the National Health Service spends £2 billion per year treating patients who have had adverse reactions to drugs prescribed by their doctors (Boseley, 2008). Recognising the problem, authorities in a number of countries have established bodies for monitoring and advising on ADRs. For example, in Australia, the Adverse Drug Reactions Advisory Committee has been advising on the safety of medicines since 1970.

As the Royal Society noted in its 2005 report on the development of personalised drugs, the current 'trial and error' approach to drug treatment which guides the choice of drug and dose has a wide range of efficacy and side effects. This is due to a number of factors, including differences in the age, weight, sex, and ethnicity of the patient; the nature of the disease; the diet of the patient; the dose of the drug; the other drugs and remedies the patient is taking; whether the drug is taken as prescribed; and potential genetic variation in response to the drug (The Royal Society, 2005: 6–8; see also Edwards and Aronson, 2000). It is believed that a better understanding of the genetic basis of disease will lead to the development of drugs that are more clinically effective and will overcome the problems associated with variability in drug responses, including ADRs which may cause death. This overlooks the fact, highlighted by recent research, that a large proportion (estimated two-thirds) of current ADRs could be avoided through improved prescribing practices, obviating the need for pharmacogenetic testing (Pirmohamed, et al., 2004, citing Hedgecoe, 2010: 173). Health authorities and research councils, however, have been keen to support a research agenda that is seen as likely to lead towards the development of pharmacogenetic medicines. The concept of personalised medicine fits comfortably with the genetic worldview (Miringoff's (1991) term) that has increasingly informed policymaking in a number of societies that pursue neoliberal policies (see Petersen and Bunton, 2002). The notion that health and behavioural problems—once seen as amenable to collectivist solutions within the welfare state—are problems residing with the individual (e.g. a genetic weakness or predisposition) that necessitate genetic-based treatments has gained widespread currency among many scientists, clinicians, and policymakers. While many of the imagined treatments are unlikely to be realised due to the complex nature of diseases, which makes the genetic contributions to particular conditions difficult if not impossible to decipher (Holtzman and Marteau, 2000), changes in institutional arrangements have been implemented on the premise that genetic tests and treatments will in the future become a routine part of healthcare practice.This move towards the 'mainstreaming' of genetics has occurred with little public discussion or meaningful consultation with the various groups that will be directly or indirectly affected by these changes in healthcare delivery. Bioethicists and other commentators have raised questions about the individual patient and family impacts of genetic technologies, and bioethics principles have been utilised in developing protocols in guiding healthcare practice. They have, for example, debated patients' and their families' 'right to know' or 'right not to know' whether they have a genetic condition (Chadwick, et al., 1997), and efforts have been made to protect individuals' autonomy in counselling following a genetic diagnosis (e.g. Clarke, 1990, 1991) (see Chapter 5). These debates have concerned clinician–patient encounters and the dilemmas confronting individuals who are undergoing genetic testing and their families, rather than broader issues pertaining to the social context and desirable social arrangements and values. Few writers have questioned the inevitability of the envisioned genetics-based healthcare future or critically examined the attendant expectations about technologies and their implications.

In the UK, the Department of Health's White Paper, *Our Inheritance, Our Future: Realising the Potential of Genetics in the NHS*, published in 2003, marks the approximate beginnings of the effort to realise such a vision, signalling changes planned in services and training to accommodate the vision of the genetics-based healthcare future. This included the training of a new generation of genetic counsellors (see Chapter 5). A review of this White Paper, published in 2008, outlined a number of 'models' adopted by the UK government to integrate genetics expertise into mainstream healthcare. These included a number of pilot projects to enhance patient access to genetics services and to bring specialist genetics advice into genetics services. The review also noted that 'significant progress' had been made on commitments to genetic screening, including for Down syndrome, sickle cell, and cystic fibrosis (House of Lords Science and Technology Committee, 2009: 39). A House of Lords report, *Genomic Medicine*, published in 2009, emphasised the need for efforts to translate research in genomic medicine into clinical practice, and for continuing capital investment to 'modernise the genetic services' and to cope with an anticipated increasing demand for genetic tests (2009: 40). In Australia, in recent years, new structures and new forms of service delivery (e.g. coordinated statewide approaches to screening for particular genetic conditions) likewise have been developed in line with expectations about the future availability of new tests and therapies (Australian Health Care Associates (n.d.)). Such initiatives, in turn, produce their own expectations. Once they are established and personnel have been trained in anticipation that more genetic tests and therapies will be developed, they lend support to the belief that innovations are indeed imminent and that health care professionals need to be prepared.

The health planners who have designed the new genetics-based health-care arrangements and the bioethicists who offer their perspectives on the proposed innovations have reflected little on whether the general direction of the changes are desirable and what the impacts are likely to be on healthcare provision, views on health and illness, and the quality of patient care. It is largely accepted by various interested constituencies that genetics will deliver what is promised and that related treatments will necessarily improve health and the standard of healthcare. The overriding challenge is seen to be how best to regulate these developments so as to minimise the deleterious impacts for the individuals who are involved and their families. Fundamental questions about the overall direction of genetics-based healthcare and the visions and assumptions which guide it—especially the promise of 'personalised medicine'—remain unaddressed.

Definitional Problems and Practical Limitations of 'Personalised' Medicine

Research on a range of personalised medicines, including drugs for Alzheimer's disease and breast cancer, has proceeded on the assumption that 'tailor-made' pharmaceuticals are not only feasible but deliverable in the foreseeable future. This is despite the fact that, to date, there have been few research breakthroughs resulting in innovations that lend support for this contention (Herceptin, a drug for breast cancer, is arguably the exception) (Hedgecoe, 2004b). The term 'personalised medicine' itself is problematic, being imprecisely defined and misleading as a description. According to the US Biotechnology Industry Organisation, personalised medicine is 'a new approach to healthcare' that 'will allow healthcare providers to identify the most appropriate therapeutic intervention and/or dosage for an individual based on his or her personal bio-molecular characteristics, thereby maximizing clinical benefit and reducing the risk of side effects' (Biotechnology Industry Organisation, 2009). While the concept of 'tailor-made' medicine has obvious appeal and is easy to convey to lay publics, fitting comfortably with dominant individualistic conceptions of health and illness, as Hedgecoe (2004b) observes, personalised medicine is not strictly 'personal' in the sense of it being shaped around an individual's specific needs; rather, it puts people into treatment groups based upon their shared genetic characteristics. Personalised medicine is at best likely to offer a spectrum of drugs that limit (although not eliminate) the potential for ADRs by targeting groups that have similar genetic traits. Further, as Davies (2006) notes, in a recent article assessing the development of personalised medicines, 'the generalizability and clinical application of pharmacogenetics has proved more challenging than expected' (2006: 111). Difficulties include limited clinical effectiveness for specific genotypes, unravelling the complexity of gene–gene interactions, and the lack of information on how genetic risk factors interact with environmental factors (Davies, 2006).

Despite the proliferation of start-up companies offering to decipher individuals' genomes for a fee, which lends the impression that the science is moving at a rapid pace, there is thus far no scientific consensus on the risk factors for the most common complex diseases such as heart disease and cancer (Ikediobi, 2009: 85). The majority of genetic risk factors that have been discovered through genome-wide association studies do not make a significant contribution to disease (Feero, et al. 2008: 1351). Many markers are yet to be discovered. And, for some known genetic markers, their contribution to disease is poorly understood (Feero, et al. 2008). In its assessment of the field of personalised medicine, the UK's Royal Society commented that, on the basis of the evidence of the complexity of genetic diseases, 'it is unlikely that there will be an immediate change in clinical practice based on pharmacogenetics' (The Royal Society, 2005: 41). It noted that any clinical applications will be gradual and that 'its true potential may not become apparent for 15–20 years' (2005: 41). Workshops organised by The Royal Society, involving members of the public, revealed a number of social and ethical concerns in relation to the translation of pharmacogenetic research into practice. While 'On balance, most participants saw the potential development of pharmacogenetic testing as beneficial in providing information to make choices about diseases affecting them and treatments available . . . a significant minority of participants were concerned about the increasing use of genetic tests in society' (The Royal Society, 2005: 39). Major concerns expressed were the protection of patient sovereignty and the role of the professional in offering impartial advice to enable people to make informed choices, and whether accompanying institutional arrangements could allow the successful delivery of the technology associated with pharmacogenetics (The Royal Society, 2005: 39). In its concluding comments, The Royal Society noted,

> In considering the delivery of pharmacogenetics, the [UK] Department of Health will have to take into consideration the public concerns and expectations of the applications of genetic technology. Our public dialogue exercise highlighted areas where the participants' views differed from those of the working group, for example where future expectations about the delivery of genetic tests by pharmacists conflict with the preferences of the public. (2005: 42)

As this report underlines, public responses will be crucial to realising advances in the field of personalised medicines. As with other technologies, public support for innovations cannot be taken for granted, and it should not be assumed that all groups in the community share the hopes and expectations of patient groups and many health care professionals (see Chapter 6). Indeed, the available evidence suggests that different groups hold different perspectives and offer varying levels of support for biotechnologies and of trust in experts. For example, The Royal Society workshops revealed that participants from Black and minority ethnic backgrounds had the lowest levels of trust in professionals (2005: 38). What support there is may evaporate if promised drugs and treatments do not materialise, or if trust in researchers and professionals is lost, for example, through adverse publicity about the side effects of particular drugs (such as happened in 2009 with the anti-arthritis drug, Vioxx) (Rout, 2009: 3) or fraudulent research practices.

The most immediate threat working against the realisation of personalised medicine, however, is the potential drying up of investment in the pharmaceutical and biotechnology sectors due to market failure. The economic crisis of 2008–2009 revealed the vulnerability of the biotechnology and pharmaceutical sectors to major shifts in market sentiment. Until 2009, both sectors grew rapidly. For the seven years until 2006, the sales of global pharmaceuticals grew 80 percent, from \$US356 billion to \$US643 billion (Walters, 2009: 23). However, both the sales and profits of pharmaceutical companies have dropped in recent years due to the expiration of the patents on the blockbuster drugs that have powered this growth and the flight of investment during the financial crisis. To add to their woes, other patents are due to expire within the next three to five years (Walters, 2009: 23). Despite a wave of recent take-overs, for example, by Merck of Schering-Plough, in the expectation that consolidation will solve their problems, there is evidence to suggest that the benefits of pharmaceutical mega-mergers will fail to materialise ('Bitter pill underlines drug giants' inability to innovate', *The Weekend Australian Financial Review*, March 14–15, 2009: 32). The biotechnology sector has also suffered badly due to a severe decline in share prices, as investors flee from the risks inherent in drug development and the long periods required for new innovations to come to the market (Walters, 2009: 23). This has affected areas such as stem cell research, which is predicted to lead to a range of therapies in the coming years (Jones, 2009) (see Chapter 4).

In short, for a range of reasons—economic, scientific, and 'social'—the ideal of personalised medicine is unlikely to be realised in practice, at least in the foreseeable future. However, many constituencies—including science groups, biotechnology and pharmaceutical industries, and patient groups—share the view that such medicines are not only feasible but imminent and work to keep the vision alive. In some cases, expectations are based on hope—as in the case of patients who are desperate for treatments—but for other groups (researchers, industry groups) economic vested interests are evident. Many groups directly or indirectly benefit financially and otherwise from the heightened expectation that new genetic innovations are 'on the horizon'. The origins of claims of genetic innovations and their significance, therefore, need to be carefully scrutinised and the underlying interests exposed.

STEM CELL RESEARCH

The field of stem cell research perhaps more than any other area of bio-technology illustrates the power of expectations to mobilise support for research. In Chapter 4, I delve into this field and its promises in detail. However, in order to underline the motivating power of expectations, it is worth examining a number of the promises here. During the early years of the twenty-first century, many countries have begun to invest heavily in research on stem cell treatments, with predictions of impending biotech 'breakthroughs' appearing in the news media almost daily. The birth of Dolly represented a crucial juncture in scientists' efforts to gain control over processes of life, in that it increased expectations that the cloning tech-niques employed could be used and in all likelihood soon would be used to develop therapies for regenerating or repairing diseased and damaged tissue. Stem cells are undifferentiated cells that have the capacity to become differentiated (to develop specialised functions) and also to give rise to more stem cells. Although growing expectations have surrounded stem cell tech-nologies, cloning techniques, and tissue engineering since the late 1990s, the origins of stem cell research and regenerative medical experiments can be traced back to the late nineteenth century with research in plants and in the early twentieth century in animals (Maienschein, 2009: 34; see also Maienschein, 2003). Through the twentieth century, several different lines of research focussing on aspects of regeneration were undertaken—guided by essentially the same questions and addressing the same problems that concern researchers today (Maienschein, 2009: 33). As Jane Maienschein, et al. argue, stem cell research 'is a poster child for translational research' (2008: 48). That is, it embodies the currently dominant ethos that research *will* produce concrete outcomes. Translational research, Maienschein, et al. note, 'has become a mantra in Washington, DC, and beyond':

> Instead of implicit promissory notes about eventual results, scientists must promise specific results up front. Moreover, they must produce results sooner rather than later and more specifically targeted for par-ticular ends rather than for general good. Finally, there is now far more guidance from public investors. The result is an ethos of translation. (2008: 43)

In the view of these authors, it was the discovery of so-called pluripo-tent stem cells in humans in 1998 and the cloning of Dolly, the sheep, in 1997 that changed thinking about the prospects of application. After these 'breakthroughs', 'the solid details of the science became much less important than the prospects—or fears—for application' (2008: 48). Fur-ther, with the prospect of the study of the development of human plu-ripotent stem cells, scientists were called upon to translate their research into biomedical applications and address political, religious, and ethical demands to limit research involving embryo research (2008: 48). However, it is recognised that there is nothing inevitable and straightforward about the process of translation and that there is a need to re-engineer health research institutions to facilitate the 'bench-to-bedside' translation (2008: 44). For example, there are designated institutes for stem cell research or regenerative research, which often include discussion of translation in their mandate (2008: 49). Examples are the California Institute for Regenerative Medicine and the Australian Stem Cell Centre (see Chapter 4). This focus on translation has changed the way science is undertaken and scientists' self-conceptions of their own work.

At Monash University, where I work, in April 2009, it was announced that $153 million would be allocated to the Australian Regenerative Medicine Institute (ARMI), with the University investing the bulk of the funds ($103 million), the Victorian Government committing $35 million for equipment and the fit out of laboratories, and the balance ($15 million) provided by the federal government. According to the project's website:

> At full capacity ARMI will be one of the world's largest regenerative medicine and stem cell research centres. Its scientists will focus on un-raveling the basic mechanisms of the regenerative process, enabling doctors to prevent, halt and reverse damage to vital organs due to disease, injury or genetic conditions. This work will form the basis of treatments for conditions such as neurodegenerative disorders, diabetes, arthritis, musculo-skeletal and cardiovascular diseases . . . (http://www.med.monash.edu.au/armi/) (Accessed 9 September 2009)

By any measure, this is a substantial investment of public funds and has been made on the promise that new treatments for a range of diseases will follow from a better understanding of the mechanisms underlying regenerative processes. However, the evidence to support claims of impending benefits is lacking and relies considerably on the confidence of publics and policymakers that researchers can deliver what is promised. This places researchers under considerable pressure to produce outcomes of demonstrable public benefit. However, a large expenditure of funds and resulting pressures do not guarantee successful outcomes and, in some cases, as was demonstrated by the Hwang case, may lead to fraudulent practice (see Gottweis and Kim, 2010; Kitzinger, 2008). A major challenge facing large-scale research programs in any field is that of governance, where mismanagement, conflicting stakeholder goals, and personality clashes between key actors can affect the dynamics of research and development. This can be especially acute where there is an expectation to produce outcomes within short time frames. Such pressure seems to have been evident in the history of the Australian Stem Cell Centre, with reports in 2008 of internal conflicts over its direction leading to the formation of an interim board to take over management of the Centre. According to a news report at the time, 'An argument about whether the centre should be pursuing commercial objectives rather than more basic research geared towards understanding stem cells was at the heart of the dispute between the centre's board and its stakeholders, which include universities and other research institutes' (Medew and Leung, 2008). It would appear that tensions had at least partly been generated by the government's high expectations about imminent applications: the previous, Liberal government had 'committed $100 million to the centre until 2011 on the basis that it would achieve some commercial objectives'. In a context of such expectations, the stakes are bound to be high, especially with the treat of discontinued funding (and truncated careers) with the failure to deliver on promises within anticipated time frames.

Critics and some scientists have expressed their concern that hype about stem cell research may raise unrealistic expectations among patients and publics about the conditions that are treatable and the timelines for applications. Fears have been raised that translation may be difficult to achieve because not enough is known about the basic science. The rush to find clinical applications may obscure the difficulties in realising the visions (Maienschein, et al., 2008: 48). The field of stem cell research is contentious especially where human embryonic stem cells (hESC) are used because it is seen to involve decisions pertaining to the significance of life. Although there are different varieties of stem cells, hESC are seen to have the greatest potential for regeneration because they are pluripotent. However, research involving these cells involves the discarding of embryos which is seen by some people to transgress rights to life. Debate in a number of countries thus has focused on 'the advantages for the many against the harm to some embryos that may or may not be considered as full lives' (Maienschein, 2003: 6). In the US, this debate has been particularly vigorous, especially in the lead up to George W. Bush's decision to allow some federal funding for research on only some stem cell lines (see Chapter 4). However, the issues raised in the US have been raised to varying degrees in many countries. In my own ongoing study of public engagement in relation to stem cells, undertaken with a colleague at my university, a number of the interviewed scientists expressed concerns about the danger of creating unrealistic expectations among patients about the timelines for developments. Some worried that desperate patients may be led to search for expensive and unproven treatments offered in countries with poorly regulated research. They also felt that the creation of what they saw as unrealistic expectations ultimately undermined confidence in stem cell research. Although news items reporting 'breakthroughs' in stem cell research appear regularly in the media, in the absence of alternative and contextual information, it is difficult for publics to assess the validity of such claims.

Direct-to-Consumer Advertising of Stem Cell Treatments

Many news reports and advertisements for stem cell-based medical treatments lend the strong impression that the science is settled and treatments currently exist. A study by Lau and his colleagues of websites that advertised treatments for a range of diseases and injuries using stem cells, involving clinics based all over the world, found that the direct-to-consumer portrayal of stem cell treatments 'is optimistic and unsupported by published evidence' (2008: 594). There is a tendency to portray therapies as 'safe and effective for a broad range of diseases in the context of routine clinical use', despite there being no clinical evidence to support the use of such therapies for the routine treatment of disease (Lau, 2008: 592). The authors suggest that providers are making inaccurate claims in their promotional materials and that clinics 'may also be contributing to public expectations that exceed what the field can reasonably achieve' (2008: 594). In a context in which the internet is becoming a major source for patient information, such websites have the potential to considerably influence people's treatment decisions. As Lau and his colleagues note, people who use such sites may not be receiving appropriate and sufficient information to enable them to make the best decisions. When one considers that the treatments can be very expensive—$US21,500, excluding travel and accommodation for patients and care givers, was the average costs on four of the websites where these were mentioned—and that serious side effects from treatment have been documented (Dobkin, et al., 2006, citing Lau, 2008: 594), the potential for exploitation is great.

The information posted on the numerous websites devoted to stem cell treatments seems persuasive, especially because many treatments are endorsed by patients and purportedly reputable clinicians. A qualitative analysis of the content of websites that advertise stem cell treatments, undertaken with a colleague (see Petersen and Seear, forthcoming), highlights the diversity of conditions subject to stem cell treatments, with some sites including accompanying claims to benefit and testimonials from patients recorded via webcam. They are sometimes hosted by patient groups/forums and include scientific information on stem cell treatments. Links to blogs and other forms of communication are common. One example is the Patient Forum of the Safe Stem Cells Movement, which includes information on stem cells, a 'breaking news' section, a 'frequently asked questions' page, 'patient clinic reviews', 'patient voices', a 'guestbook' (with dated entries from individuals (including patients) expressing their support for stem cell research), and links to Facebook and Twitter (http://www.safestemcells.org/Home_Page.html) (accessed 18 September 2009). Some sites advertise the nature of the treatments that are available, often including patient testimonials, news items, and contact details. In many cases, it is difficult, if not impossible, to determine who is sponsoring the website and to gauge whether the individuals or organizations mentioned are reputable. The website titled 'Returning Hope' includes a logo on its front page, 'Bangkok Hospital Medical Center', but interestingly includes a link, 'Search for a hospital', which lists various hospitals offering a range of services. All are described in terms of their excellence or 'technological sophistication': 'one of the most technologically sophisticated hospitals in the world today' (Bangkok Hospital); 'This hospital is an example of world class medical facilities located throughout Asia, Doctors are American Board Certified and the hospital has been profiled on CBS News, 60 minutes' (Bumrungrad International—World Class Hospital in Thailand); 'Piyavate hospital's vision includes providing it's doctors with modern medical technologies and training in the latest, hi-tech medical procedures & medicine while continuing to provide it's services to patients at competitive prices to encourage the trend of medical tourism to Thailand' (Piyavate Hospital). It is not clear why these particular hospitals are being recommended and what their connection is. This website also includes 'Recent patient testimonials', including webcam recordings of patients who suffer a range of conditions, including diabetes, stroke, ataxia, and autism. Such information is framed in a way that seems intended to engender confidence in the named organisations, practitioners, and technologies. There is nothing to suggest that the basic science underlying stem cell treatments is contentious, that treatments have uncertain outcomes, and that risks may be involved.

A typical website, XCell-Centre, lists details on a range of therapies on offer. It notes:

> Therapy focuses on the treatment of diabetes mellitus (types 1 and 2 as well as sequelae) and stroke. Further indications comprise neurological diseases, in particular spinal injuries, multiple sclerosis (MS), amyotrophic lateral sclerosis (ALS), Parkinson's and Alzheimer's disease as well as arthritis, heart disease, eye disease, neuropathy and incontinence. (http://www.xcell-center.com/) (accessed 18 September 2009)

The website lists members of its Advisory Board, a 'News' section, and details on its physicians and laboratories (e.g. 'We cooperate with a state-of-the-art laboratory that complies with the German law on Good Manufacturing Practice'), and it offers assurances about safety of treatments and the Centre's commitment to research. For example, it notes that 'Since the start in January 2007, **more than 1600 patients** have safely undergone our various stem cell treatments' (bold in original). And, 'Beside therapeutic applications that take place at the institute, the XCell-Center also spearheads research on the medical use of adult stem cells. Two phase II studies are submitted and are scheduled to start in the second half of 2008'. Such information and assurances help convey the impression that the Centre is reputable and well-managed and can be trusted. Another website, Cell Medicine, announces that 'Stem Cell Therapy Is Available Now Outside the US and Canada', which is obviously oriented to its North American audience who may feel constrained by regulations operating in those countries. It includes a protective note that such treatment 'is not covered by most insurance'. The website includes information on a range of conditions, promising the treatment is available 'with your own stem cells'. The website has links to videos, which outline 'stem cell types and sources' and information on particular conditions such as cerebral palsy, diabetes, heart disease, kidney failure, and liver failure. Under the heading 'Research', it also includes detailed information on stem cell applications for particular conditions, with numerous references made to research evidence cited in academic journals and conference papers.

As with the XCell-Centre website, such information here performs a rhetorical function, conveying the impression that treatments are based on rigorous research and that those performing the treatments can be trusted. This impression is reinforced by patient testimonials, which confirm the benefits of treatments. The main web pages contain little information challenging or qualifying these positive portrayals. Those navigating the website have to click to the 'patient application' link to encounter the disclaimers and more qualified comments. Before the patient signs up, they are forewarned that 'The science of treatment with adult stem cells is in its infancy', 'The treatments described on cellmedicine.com are not approved by the US FDA and are not considered to be standard of care for any condition or disease', and 'For most diseases no prospective, randomized clinical trials of adult stem cells have been performed, therefore no guarantee of safety or effectiveness is made or implied' (http://www.cellmedicine.com/application/) (accessed 18 September 2009).

Patients and families who are desperate for treatments and perhaps put great store in expert advice are arguably vulnerable to accepting the claims that these websites make. In most cases, they would have no way of assessing the veracity of the claims or the qualifications or integrity of those who are making them. While it is difficult to know how users of websites respond to web-based information (this is the subject of my ongoing research), it would seem that in the context of unregulated treatments and an unregulated communication environment, individuals are prone to exploitation by disreputable individuals and organisations. The rise of the web and linked electronic communications (e.g. Facebook, MySpace, blogs, twitter), combined with the contemporary emphasis on individual autonomy in health and healthcare, has contributed to creating the ideal conditions for 'direct-to-consumer' advertising and patient exploitation. No longer constrained by time, place, and regulatory hurdles, the marketers of new stem cell treatments can readily create a demand for their services while patients can easily find a treatment among the numerous therapies on offer. The main or only constraint for the patient and their family is financial which, as mentioned, may be considerable, especially if it involves travel to another country. The long term health costs, of course, can be substantial and the risks high given that most treatments are unproven and hence not subject to regulatory oversight.

SOURCES OF EXPECTATIONS

Although in many cases it proves difficult to disentangle the various contributions to the generation of expectations, scientists are likely to play a major role. Scientists' established role as news sources on science 'breakthroughs' strategically positions them to establish the agenda for debate on pertinent issues (Conrad, 1999; Petersen, 2001). Scientists often emphasise the benefits of their research, especially during the early phases of the development of a new field, in order to solicit the interest and support of funders. The following articles on personalised medicine, which appeared in two Australian newspapers and a UK newspaper, respectively, in recent years, are not untypical of news reporting in this particular field:

Genetic screening and privacy

We are only a few years from the threshold of what has been called 'personalised medicine'. This is when individual DNA sequences will be cheaply and readily available. When this occurs, medical research will address itself increasingly to the challenge of producing medicines that target those genes an individual has that are likely to cause disease. . . .
(*Sydney Morning Herald*, 12 March 2009, p.1)

Dawn of personalised medicine

When Steven Roebuck booked an appointment with clinical immunologist Simon Mallal in 1999, he had no idea that he would become part of a revolution in medicine. As one of a group of now over 1400 HIV-positive Australians, Roebuck's blood, time and commitment would lead to the first hard evidence that personalised medicine—treatment tailored to a patient's individual genetic make-up—was possible, affordable and, above all, useful in the doctor's surgery. . . .
(*The Australian*, 15 March 2008)

Cancer treatments will be tailored to patient's genes

Doctors have taken the first steps towards identifying genetic differences between cancer patients so that treatments can be tailored towards a person's genetic make-up.
(*The Independent*, 27 April 2004)

These stories convey the imminence of breakthroughs, such as in the reference to 'we are only a few years from the threshold', in the first story, 'Dawn of personalised medicine', in the title of the second, and 'Doctors have taken the first steps towards . . . ', in the third. Such stories rarely convey doubt or offer the qualifications about the feasibility or imminence of innovations that one might find in the more considered assessments of this field (see below). A number of studies focusing on news media portrayals of genetics and medicine confirm the generally positive depiction of genetic research; the tendency to over-emphasise the benefits of 'breakthroughs', and to downplay or overlook the risks and normative implications (e.g. Bubela and Caulfield, 2004; Nisbet and Lewenstein, 2002; Petersen, 2001, 2002; Petersen, et al., 2005). Many news reports focus on the personal and public benefits that will accrue from basic research; for example, treatments for Alzheimer's disease, heart disease, or stroke.

News reporting on stem cell research reveals a similar positive portrayal of issues, with numerous stories of imminent breakthroughs and future cures, as revealed by research undertaken in the UK and the US (e.g. Jensen, 2008; Kitzinger, 2008). As Kitzinger (2008) notes, the Hwang scandal, involving the revelation of scientific fraud, challenged this narrative, which scientists needed to reassert. This episode powerfully underlined the performative dimension of expectations that surround emergent technologies; the fact that scientists needed to 'work hard to rescue hope' which has been severely tested by the episode of fraud (Kitzinger, 2008: 427–428). Stories of promise and hope can be seen clearly in the following articles on stem cell 'breakthroughs' published in recent years.

Europeans announce pioneering surgery

> PARIS—Physicians at four European universities have successfully transplanted a human windpipe, using stem cells from the recipient's own bone marrow to reline a donor trachea and prevent its rejection by her immune system, according to an article in the British medical journal The Lancet.
>
> The operation, performed in June, was the first to use stem cells in transplanting an airway, and is considered an important advance because it allowed the surgeons to replace a larger segment than had generally been possible in the past.
>
> (*The New York Times*, 19 November 2008)

Foetal stem cells offer hope for stroke victims

> British scientists hope to repair the damaged brains of stroke patients using stem cells from aborted foetuses, it was reported today.
>
> The UK biotech company ReNeuron is seeking permission for trials from the US Food and Drug Administration (FDA), which regulates American research.
>
> They will involve taking stem cells from the developing brain area of a 12 week aborted foetus and implanting them into patients, according to an exclusive report from the BBC. . . .
>
> (*Press Association National Newswire*, 5 December 2006)

Hope stem cells will help heart patients

> SUFFERERS of Australia's number one killer—cardiovascular disease—are being offered new hope of treatment by world-first stem cell research in Adelaide.
>
> The purified stem cell technology is showing early promise and will give sufferers a new treatment option using cells harvested from their own bone marrow.
>
> The aim is to help regenerate damaged heart tissue. While it is not designed to supersede current treatment, it gives doctors another option for patients with heart failure who are not responding to standard management, such as medication, surgery and pacemakers. . . .
>
> (*The Advertiser*, 11 July 2006)

News coverage of cases involving high-profile (celebrity) individuals and families in search of stem cell treatments has given considerable prominence to this field in recent years. Of particular note is the case of the US actor Christopher Reeve, a strong advocate of stem cell therapies who looked forward to a cure for his paralysis from a spinal cord injury resulting from a horse riding accident. Extensive news coverage of his plight, along with research, lobbying, and communication efforts associated with the organisation he and his wife established, the Christopher and Dana Reeve Foundation, have undoubtedly helped give prominence to the potential of stem cell therapies (see http://www.christopherreeve.org/site/c.ddJFKRNoFiG/b.4409743/k.C825/About_Us.htm) (accessed 25 September 2009). In the UK, the potential benefits of stem cell treatments have also been given prominence through some highly publicised cases, such as the Whitaker and Hashmi families' efforts to use stem cells from 'perfectly matched siblings' for the treatment of their diseased children (Petersen, et al., 2005).

Bubela and Caulfield's (2004) research emphasises that scientists may not always be the source of strong or exaggerated claims about science 'breakthroughs' that are reported in the media. Elsewhere, Caulfield suggests that commercial interests may bias reporting towards an uncritical portrayal of issues (2004). He refers to evidence showing that articles in peer-reviewed journals are more likely to report the benefits when the research is funded by industry (2004: 178). On the other hand, negative results are either de-emphasised or not published at all. He concludes that 'Commercial influence on public representations of science has the potential to create a skewed picture of biomedical research—a picture that emphasises benefits over risks, and predictions of unrealistic breakthroughs over a tempered explanation of the incremental nature of the advancement of scientific knowledge' (2004: 178). Caulfield argues that the trend towards positive, industry-influenced reporting may serve the same purpose as an explicit promotional campaign. In fact, he suggests that it may operate more powerfully than an explicit promotional campaign in that the message is separated from an obvious promotional agenda and is often connected to a trusted person such as a university-based researcher (2004: 178–179).

While news articles do not generally note the nature and level of industry support for research that is reported, examination of the websites of some researchers whose work is positively reported reveals strong industry links. A study by GeneWatch UK into press reports on research into genetic predisposition to smoking-related lung cancer, for example, revealed that the corresponding author had a history of substantial funding from the tobacco industry (GeneWatch UK, 2009: 23, 52, 88; see also GeneWatch, 2010a). The GeneWatch report cites subsequent research showing that individuals exhibiting the genetic variant do not carry a significantly increased risk of lung cancer (GeneWatch UK, 2010b: 58). However, such disconfirming research has not been reported. As noted earlier (p. 26) it is easy to see why the tobacco industry would favour the view that some individuals are genetically susceptible to lung cancer because it suggests that the rest of the population can 'smoke with impunity' (GeneWatch UK, 2009: 58).

The Role of Public Relations

A point that is often not recognised by media analysts is that many news stories in science and other fields are generated by public relations sources—which may be funded by industry, government, and non-government organisations—whose main aim is to generate a 'positive spin' on developments. The estimates of the proportion of stories generated by public relations sources vary between studies and across time. As Jim Macnamara points out, the influence of public relations on news stories can be difficult to ascertain because 'public relations' may include a vast array of activities, of which media relations is but part, and it sometimes proves difficult to ascertain the source of stories (2009: 11–12). Further, journalists may incorporate public relations professionals into their circle of contacts as 'trusted sources' and no longer see them as 'public relations' (2009: 10). They may be specialists with long backgrounds in particular industries (e.g. IT, finance, motoring, fashion) and are seen as such by journalists. However, data from quantitative data conducted over more than 80 years suggests that 'somewhere between 30 and 80 percent of media content is sourced from or significantly influenced by public relations practitioners, with estimates of 40–75 per cent common' (Macnamara, 2009: 8). While Macnamara provides no breakdown for science news content, it is likely that PR influence is at the higher end of his estimates. This is because a large proportion of science news is source generated and is released by universities, research institutes, and the like, which are reliant on research funding and are keen to promote a positive image of science.

Many years ago, Dorothy Nelkin (1987) noted the crucial role of public relations in science communication which, she argued, is a means by which scientists seek to control the media agenda by offering a positive portrayal of their work. As Nelkin pointed out, the public relations industry has played a strong role in science since the mid-nineteenth century. However, after World War I, the expanding scientific enterprise and the need for greater funding brought greater pressure from scientists to enhance their public image. Public relations activities by scientists increased again after World War II, in the wake of the 'space race' and burgeoning research in the biosciences. By the 1960s and 1970s, public relations activities had become integral to the operations of professional societies, academic institutions, and research organisations, and their members began to receive instructions on how to deal with the media. Nelkin notes, for instance, that the American Institute of Physics expanded its publicity programs in the 1960s, offering seminars for journalists and news conferences to outline newsworthy developments in physics (1987: 137). The organisation subsequently sent out routine press releases and offered instructions to physicists on how to deal with reporters. In the UK, the premier science body, The Royal Society, has run communication skills and media training courses for scientists for a number of years to equip them to engage with 'a variety of audiences, whether it be students and scientists from other disciplines, or the media and mass public' (http://royalsociety.org/page. asp?id=1151) (accessed 4 September 2009). The Royal Society's website also features regular 'Science in the news' bulletins to inform publics about important science developments. In Australia, the science body, The Federation of Scientific and Technological Societies, which represents 60,000 Australian scientists and technologists, runs an active science communication program. This includes regular media releases on science issues and a 'Science meets Parliament' event, which brings 'up to 200 scientists from all over the country for face-to-face meetings and forums with Parliamentarians in Canberra' and 'allow the scientists unparalleled opportunities to witness national decision making at first hand, and to inform this process on important scientific issues' (http://www.fasts.org/index.php?option=com_content&task=view&id=28) (accessed 4 September 2009).

There is evidence that scientists are becoming more strategic in the use of media for the purposes of self-promotion and more willing to talk to journalists and to make themselves available as sources than in the past. Indeed, it has been suggested that PR-related goals of media engagement may be overshadowing other goals, such as the communication of quality information to publics (Peters, et al., 2008). Staged news events and news releases provide a key means for scientists to control the flow of information to publics and for deliberately or inadvertently heightening expectations about innovations. Gina Kolata (1997), who has a background as a science writer, has outlined the significance of news releases in the breaking of the story of Dolly, the cloned sheep, in 1997. As she argues, science journals such as *Nature*, which published the article on Dolly, seek to gain maximum publicity for their articles by sending journalists 'tip sheets' containing short descriptions of its articles for the forthcoming issue. These arrive in the reporters' e-mail boxes a week before the journal's publication, and the next day reporters are faxed papers. In return for these 'tips', journalists undertake not to publish or broadcast stories until the papers are published. In this case, the story was broken by a journalist from the UK's *The Observer*, who obtained his information from sources other than *Nature*, thereby avoiding breaking the embargo (1998: 26–30). The rise of e-mail and the World Wide Web has provided new means for scientists to control representations of science, to attract attention to their research, and to engender a positive portrayal of their work. However, a study of the content and language of the web pages of German universities and non-university-based research institutions suggests that they may not always be designed with a clear target group (e.g. journalists) in mind (Lederbogen and Trebbe, 2003).

There is little doubt that scientists' increasing reliance on funding through biotechnology and pharmaceutical companies will predispose them to 'hype' the practical significance of their research. Company share-holders need to be persuaded that investments in research will pay dividends through the development of new drugs. In the wake of the mapping of the human genome, considerable expectations surround pharmacogenetics, a field of research which aims to ascertain how people's genetic makeup affects their response to medicines (Hedgecoe, 2004b). Pharmacogenetics accepts the fact that individuals vary in their responses to drugs: the way they metabolise drugs, the ways drugs operate in their bodies, and the rate and extent to which products are removed from bodies (Hedgecoe, 2004b). Pharmacogenetics marries two substantial communities of interest: the pharmaceutical industry, on the one hand, and the genetics community, comprising the biotech industry, genetic researchers, clinicians, genetic counsellors, and patient groups, on the other. Given this overlap of interests, it is hardly surprising that personalised medicine has achieved such strong political support. These interests indeed often work collaboratively to educate the public and lobby for more research. For example, the Personalized Medicine Coalition, a self-proclaimed independent, non-profit group, states on its website that it 'educates federal and state policymakers and private sector healthcare leaders about personalized medicine, helping them understand the science, the issues and what is needed for the positive evolution of personalized medicine' (Personalised Medicine Coalition website: http://www.personalizedmedicinecoalition.org/about/mission-goals.php) (accessed 15 April 2009). Groups such as this constitute part of the large and growing PR 'industry' that dominates the public representations of genetics in the increasingly diverse, including online media. The online environment has allowed the promoters of genetic innovations a ready means to reach a broad audience, including patient groups that have come to constitute a new form of sociality, 'biosociality' (Rabinow, 1992), unimaginable in the pre-web era.

The power of expectations to mobilise actors in the field of genetics is apparent in the high level of science and policy support for 'biobanks'—population-based databases that comprise genetic, medical, and lifestyle information—that have emerged in many countries in recent years. Biobanks are ascribed a key role in 'unravelling' the genetic, lifestyle, and environmental contributions to disease (see Chapter 3). Governments and funding bodies have supported these initiatives directly through their funding, and bioethicists, social scientists, and regulators have assisted in the effort to develop a social and regulatory environment conducive to their development (Gottweis and Petersen, 2008). Governments have placed great store in their potential economic benefits, and, like other biotech innovations, biobanks have been closely linked to the project of nation-building (Jasanoff, 2005). Although diverse in their size, operations, timelines, and modes of governance, biobanks share the fundamental aim of advancing understanding of the genetic contributions to disease. While genetic databases have existed in different forms for over thirty years, in recent years, a new generation of biobanks has emerged that are large scale (population-wide) (sometimes, as with Iceland, including the entire population) and prospective, sometimes spanning many decades into the future (see chapter 3). Given their scale and time frame, biobanks call for a large element of faith from publics, funders, and policymakers that they will eventually deliver what is promised, namely, the breakthroughs that will lead to the new generation of personalised medicines. They provide an exemplar of 'big science' in action, seen also with stem cell research, in that they are conceived as large infrastructure projects that call for considerable investment and cooperation between researchers and research teams at the global level (Gottweis and Petersen, 2008). The different kinds and levels of support and inputs required for their operation and sustainability over a long period of time also make them vulnerable to changing politico-economic dynamics and policy priorities.

The ethical and governance challenges of biobanks will be discussed in more detail in Chapter 3. However, in brief, these include the ongoing difficulty of achieving legitimation in the face of substantial criticisms by legislators, scientists, and others of some projects (e.g. UK Biobank), including in relation to their scientific, commercial, and public benefit aspects; problems in sustaining funding over the longer term; and expressed public and NGO concerns about industry links, loss of privacy, and the commodification of the body. A number of biobank projects have encountered difficulties during the early phase of their development, with some running into considerable problems and being effectively abandoned, as in the case of Iceland's Health Sector Database) (Pálsson, 2008). In late 2008, this project's industry partner, De Code, ran into difficulties, with the collapse of its shares, stiff competition to its gene testing business, and a cash flow problem (Langreth, 2008). As this and other experiences of biobanks reveal, the generation of consistently high expectations can be difficult to sustain, especially when economic conditions change. However, as I will explain in Chapter 3, despite such failures, expectations continue to provide a considerable motivating force for innovations, as is illustrated by the case of UK Biobank.

QUESTIONING EXPECTATIONS

Despite mounting evidence of the limitations and adverse implications of the genetic conception of health and of overblown expectations of stem cell therapies, bioethics has failed to offer a systematic analysis of the promises of these fields. 'Big picture' questions about where these fields are heading, who benefits or is likely to benefit, and who is disadvantaged or likely to be disadvantaged remain unaddressed. There has been little questioning of the premises that guide research priorities and research funding decisions. In relation to genetics research, the significance of politico-economic issues and interest group influence, especially industry influence, outlined above, has been largely overlooked in ethical debates about personalised medicine thus far, which has largely focused on questions concerning autonomy, consent, protection of privacy, access, 'rights to know', and 'benefit sharing'. With regard to stem cell research, there has been little attempt to interrogate the industry interests and science-commercial links shaping priorities for research and development in the field. Debates have focused largely on the moral status of the embryo and the rights of women (see Chapter 4). There has been relatively little focus on the broad justice implications of envisaged developments—for example, the inequitable impacts on the community of the allocation of considerable resources to this field and the significance of heightened expectations for patients and their families who are desperate for treatments. Reference to bioethics' principles and language (autonomy, rights, moral status, and so on) has served to 'narrow' the debate on the substantive questions raised by genetic developments (see, e.g. Evans, 2002), such as those discussed above, while lending a veneer of legitimacy to decisions. There has been little effort to address the 'democratic deficit' that excludes the majority of the population from decisions affecting the development of technologies.

An understanding of the dynamics of expectations can inform debates and decisions about future technologies. The costs of high expectations, especially in skewing funding towards innovations that may benefit only some groups, need to be acknowledged. Money invested in R&D that is unlikely to be realised or to be realised only many years away, and then may benefit only certain people, may be better spent on services or activities that improve the health and wellbeing of whole communities; for example, services to assist people to manage their conditions at home and in the community. The history of biotechnology R&D shows that access to new treatments tends to be unequal in the population (Law, 2005; Moynihan and Cassels, 2005). While there are 'winners' and 'losers' with any new technology, the social distribution of benefits of biotechnologies is likely to be especially uneven given the high costs of taking treatments to the market. Pharmaceutical companies are driven primarily by the profit motive not patient welfare, and the system of drug patents works to keep costs of drugs high and thus inaccessible to those who do not have the means to pay. Further, given budgetary constraints, governments cannot be relied on to redress inequalities of access to new treatments. Policies that affect access to drugs, such as the system of reimbursement for medications (for example, under Australia's Pharmaceutical Benefits Scheme), may become restrictive, especially in a context of burgeoning healthcare costs (see, e.g. Sharp, 2009), thus disadvantaging poorer groups whose access to drugs is limited to the public health system. These unequal impacts of future drug development are rarely part of public discussion about promising new biotechnologies, but should be so that publics are better able to make informed decisions when assessing claims about particular innovations.

Making more modest, qualified claims about biotechnologies is essential to the establishment of public confidence in science and technology in general and to creating the conditions for wider community engagement in technological decision-making which should be central to the democratic process. As scientists and policymakers increasingly recognise, public trust is an important precondition for research and development and has implications for people's engagements with technologies. In the UK, scientists and policymakers have expressed concern that a 'crisis of trust', following in the wake of a series of events, including the BSE fiasco, the GM controversy, and the GM Nation debate and various medical scandals, can lead to public disengagement from and resistance to science and technology (House of Lords Select Committee on Science and Technology, 2000). It may lead to technologies being rejected or abused. Disengagement may also happen when biotechnology innovations, such as personalised medicines, fail to materialise in expected ways or within envisaged time frames. This may mean that future research that has a demonstrably high likelihood of improving people's lives may never gain the level of community support necessary to achieve the commitment of governments and funders. Recognising the implications of public trust for the development of science and technology, the UK's House of Lords has called for improved means of public engagement with science, 'improved communication about uncertainty and risk', and 'changing the culture of policy-making so that it becomes normal to bring science and the public into dialogue about new developments at an early stage' (House of Lords Select Committee on Science and Technology, 2000: 1.19). In the UK, recent efforts at public engagement and dialogue, for example, through the Sciencewise program (Sciencewise Expert Resource Centre: http://www.sciencewise-erc.org.uk/cms/priority-areas/) (accessed 16 April 2009), purport to set a new direction for science communication and dialogue, including in relation to biotechnologies. However, many efforts at engagement arguably perpetuate rather than challenge the so-called deficit model of public understanding that has underpinned science communication efforts in the past (Wynne, 2006) (see Chapter 6). Part of the public discussion about biotechnology R&D should address the complexities and uncertainties of science and the process of innovation. This means acknowledging the various factors that may impede the realisation of technologies and debating the implications of unfulfilled expectations which arise for a range of reasons, many of which cannot be foreseen. Such an approach seems a more honest and just way to proceed than promoting unrealistic and thus likely unrealisable expectations.

3 Engendering Consent
Bioethics and Biobanks

The recent development of biobanks—large scale databases comprising genetic, personal medical, and lifestyle information—has brought into sharp relief the limitations of the established approaches of bioethics. In particular, it has highlighted how adherence to abstract principles and the concern with a restricted range of issues has constrained debate on the substantive social, economic, and political questions raised by such collections. In this chapter, I examine how bioethics' concepts and principles have been deployed to help resolve the challenges posed by biobanks, in particular to lend legitimacy to and achieve consent for projects for which there are high expectations and considerable political support. At the outset, it is important to acknowledge that biobanks have been defined in various ways and have taken somewhat different forms in the different countries in which they have emerged. Their scale, operation, regulatory mechanisms, and, to some extent, stated objectives vary across jurisdictions. Responses to their emergence have also been far from uniform, reflecting local histories, cultures, and politics. This presents a challenge and some dangers when seeking to generalise about their nature, operation, regulation, and implications. However, since the late 1990s, there has been an emerging international consensus among scientists and influential decision makers about the value of these collections and the need to support them.

Some commentators challenge the claim that biobanks are novel. As they rightly point out, collections of genetic and medical information derived from samples of selected human populations of various sizes have existed for a number of decades in some societies, although they have generally not been called 'biobanks' (Tutton and Corrigan, 2004). For example, in the UK, the Avon Longitudinal Study of Parents and Children (ALSPAC) has collected genetic and environmental information on children and their families since 1991. Hence, in one sense, biobanks are not as novel as is often claimed. However, in recent years, a new generation of *population-wide* collections, comprising genetic information, personal medical history, and lifestyle data, have emerged in many countries, with the aim to study the contributions of genetics and environment to the development of disease over time, generally decades. It is these collections that tend to be described as 'biobanks' and that have been the focus of much recent interest among bioethicists, other scholars, policymakers, and regulators. Their *scale* and *long-term prospective nature* is seen to pose unique problems of an ethical nature in need of resolution. These new generation databases tend to be conceived as large national infrastructure projects that are seen to have potentially huge health and economic 'spinoffs'. They are envisaged as an important part of the emergent bio-economy, based on the commercialisation of biological information; on the 'translation' of research into new, marketable (and potentially highly profitable) treatments. Hence, many governments are committed to their development.

Because biobanks are prospective in nature (that is, they involve the study of diseases as they emerge, rather than after they have emerged) and necessitate the support of policymakers, commercial interests, and the community over a long period of time, they face a particular challenge in establishing and maintaining legitimacy and consent. Another unique feature is that the potential for sharing information and for combining data from different biobanks has arisen, enabled by new developments in information and communication technologies. Discussions about the 'harmonisation' of knowledge and 'transfer of knowledge' between different projects are underway, particularly within the P3G (Public Population Project in Genomics) Consortium (http://www.p3gobservatory.org/) (accessed 16 October 2009). This presents the possibility of a global research endeavour, involving different national populations, and the emergence of new forms of surveillance and regulation of populations—although this issue has hardly been discussed among bioethicists and other scholars who are concerned about the implications of these projects.

The emergence of biobanks reflects the high optimism attached to biotechnology innovations in general in recent years, outlined in the last chapter. Most biobanks are not limited to the study of particular conditions but rather allow for the study of a range of conditions that are seen to have or potentially have a genetic basis. Because in most cases they are limited neither to a particular condition nor a time period, they present particular problems of governance, as will be explained. It is envisaged that over time participants will develop diseases that can be studied for their genetic and environmental contributions. The announcement of the completion of the mapping of the human genome in 2000 created the expectation that science will in time reveal how genes contribute to disease. This expectation underpins the new research field of 'functional genomics'. As most scientists acknowledge, in most instances disease results from the contribution of a complex interaction between different genes and between genes and environments (see Chapter 2). Long-term exposure to chemicals, pollutants and ultraviolet radiation, smoking, excessive alcohol consumption, and a vast array of other physical environmental and lifestyle factors, individually and in combination, are likely to play a significant role in the onset of disease and, in some cases, may override the contributions of genetic factors. At the international level, the significance of physical environmental factors for health status is indisputable. In many poorer, developing countries, infectious diseases (often waterborne)—frequently related to a poor standard of living (substandard housing, inadequate diets, lack of clean water, and low income)—play a major role in the high mortality rates. Access to the basic necessities of life that shape health and wellbeing is a human right—which is emphasised in the Universal Declaration of Human Rights (Farmer, 2003: 196–212)—acknowledgement of which is absent in the public discourse on the ethical implications of biobanks. Of course, some genes have what is called 'high penetrance', as is the case with those affected by Huntington disease, where those who have the gene have a high probability of developing the disease. However, for the majority of conditions in most parts of the world, it is now recognised that environment plays a significant contributing role and, in most cases, a far greater role than genetic factors.

Despite overwhelming evidence of the significance of environmental contributions to disease, belief in the value of genetic information that underpins biobank projects has proved resistant to change. Biobanks, like other biomedical research innovations, are premised on a (reductionist) conception of the body as a machine comprising interlinked parts with inherent weaknesses or 'faults' (such as 'defective' genes) and are thus subject to 'breakdown' and in need of 'repair'. In line with this conception, the aim of research is to learn more about the makeup and workings of the body so that it can be 'fixed' through specialist, gene-based therapies. Recent moves to 'harmonise' biobank developments and to share data, enabled through advancements in digital technologies, is taking this reductive, fragmented view of the body and its health and functioning to a new level. The inadequacy and dangers of this conception of the body has become increasingly evident—at least to those who work outside biomedical science.

Many biobank projects were proposed or developed in the wake of the mapping of the human genome, with predictions that new breakthroughs lay just 'around the corner' with resulting economic benefits through improved healthcare. Governments see new treatments as promising to help reduce health care costs, especially those seen to be associated with rapidly aging populations, and assist in the effort to create healthy, productive populations. Biobanks are widely perceived to hold the promise of major research breakthroughs that will enable profitable new drugs, gene therapies, and diagnostic tests to be developed, establishing the basis for the envisaged bio-economy. They are seen as a kind a laboratory for developing 'personalised medicine', which many commentators predict will characterise healthcare systems of the future (see Chapter 2). The rhetoric of nation building is strongly evident with many biobank projects, which are often represented as 'flagship' scientific developments on project websites and in the professional literature. In the majority of biobank projects, the necessity for private sector involvement, by way of access to collections for the purposes of research, is assumed at the outset, with some projects, such as UK Biobank (discussed below), intending to make its collections accessible to researchers with costs calculated on a sliding scale according to the required level of access to the data. As we shall see, this reliance on industry has presented its own set of challenges for promoters and custodians of biobanks, including questions of ownership of and access to data, commercial profit versus public benefit, and the viability of projects over the long term. Again, bioethicists have remained silent in relation to many of these issues, although they touch on matters of avowed bioethical interest such as 'the public good' and 'social justice'.

THE RISE OF THE BIOBANK PHENOMENON

The rise of the biobank phenomenon and the substantial challenges biobanks pose for established conceptions of bioethics cannot be properly understood without an understanding of the profound economic, political, and socio-cultural changes under way in many societies. Biobanks are reliant on a number of predisposing conditions, including: the predominance of genetic determinism and the belief that health and social problems are amenable to solution through genetic means ('genetic welfare') (Miringoff, 1991); the existence of strong expectations among scientists, policymakers, and publics that new genetic therapeutic innovations will follow from the mapping of the human genome; the establishment of a consensus among authorities of different political persuasion of the urgent need to reign in burgeoning healthcare budgets, associated in particular with expensive technological innovations, demographic shifts (chronic older age-related illnesses and impairments) and ever rising public demands for access to healthcare, supported in large measure by 'neo-liberal' policies and practices reinforcing individual responsibility for health ('care of the self'); widespread belief among key decision makers in rational administrative solutions to emergent socio-political problems; new research opportunities created by the rapid development and merging of different technologies (e.g. digital and genetic); substantial ongoing private investment in genetic research; and the achievement of a broad consensus among publics and policymakers about the value of population-based research endeavours of this kind. Consequently, a great many groups and publics who are affected or who are potentially affected by innovations have an interest and stake in such research.

The growing liberalisation of markets from the 1980s and the increasing flows of capital, labour, and resources within the burgeoning global marketplace of ideas, products, and services are important features of the context within which biobanks have emerged and (with some exceptions—see below) flourished. Emerging biotech companies have been in search of promising new areas of investment and have often looked to 'emerging markets', such as those in Eastern Europe (e.g. Estonia) and Asia (e.g. Taiwan,

Korea, China). Stem cell research has been subject to similar expectations, with emerging markets again often being the focus of corporate interest (see Chapter 4). The promise of new genetic tests that may be marketed to the 'genetically susceptible' (i.e. those with a family history of a particular condition) or whole populations has underpinned much commercial investment (see Chapter 5). According to the World Health Organization (WHO), genetics research has considerable potential for addressing health problems at the international level, such as infectious diseases and water-borne diseases (2002). Despite acknowledging the 'different problems faced by many developing countries', such as dysfunctional healthcare systems due to the lack of 'basic necessities such as hygiene and public health' (2002: 71), the WHO nevertheless sees considerable potential offered by the genomics revolution. This overlooks that most health problems are ones resulting from poor material conditions and socio-economic disparities that are preventable through changes in social structures, institutions, and practices (Farmer, 2003).

Projects' heavy reliance on the commercial sector, however, has made projects highly vulnerable to the dynamics of the market. A number of biotech companies have jumped on the 'biobank bandwagon', only later discovering that the economic benefits are unlikely to accrue within timelines acceptable to shareholders. Some companies, such as EGeen which was involved in the Estonian biobank, retreated within a year of the commencement of the project when questions arose about the commercial viability of the venture. In this case, the company sought to concentrate on specific disease groups such as hypertension because they offered the best commercial prospects. The project consequently became a public project which, although seeking to establish a viable economic basis through international collaborative opportunities, encountered a series of funding problems and almost collapsed before finally undergoing reorganisation as an institution of the University of Tartu and receiving government guaranteed funding (with some reluctance) in February 2007 (Eensaar, 2008: 62–65).

In Iceland, deCODE genetics, the company established in 1996, by a scientist, Icelandic Kári Stefánsson, and his collaborator, Jeff Gulcher, had a chequered history from the outset until its collapse in 2009. (The company was subsequently reincarnated, with the 'new deCODE' made possible through a partnership (led by Stefánsson and Earl Collier, previously Vice President of Genzyme Corp) with drug companies to translate their research into medical therapies, rather than to create drugs in-house as before) (see Hyder, 2010). As Pálsson (2008) explains, this company was established in national and international contexts which appeared highly conducive to a large scale commercial genetics research venture of this kind. The local and international investment and socio-cultural contexts were strongly aligned. It was envisaged that data derived from the relatively homogenous Icelandic population and the excellent local historical records would be used to advance biomedical research on the genetics of common diseases. However, as Pálsson argues, despite at least initial widespread public support for the project, it never developed as envisaged due to a range of reasons. These included opposition from the Icelandic Medical Association who saw the project as violating the relationship of trust between the physician and patient, a reduction in the sample due to a growing number of people opting out of the project, and a growing backlash against bio-industries arising in part as a consequence of debates about the implications of human genome projects (2008: 47–50). In 2009, deCODE genetics filed for bankruptcy due to, according to reports, its failure to profit from translating its genetic findings into wide-scale drug production (Stone, 2009).

The advancement of information and communication technologies has been a further crucial precondition for the development of biobanks. New digital technologies have allowed new means for storing, manipulating (e.g. aggregating), and exchanging data, including genetic data, while the concurrent development of genetic and digital technologies has created the conditions for the rise of genetic epidemiology—a relatively new discipline oriented to the 'mapping' of genetic disposition among populations—that has been a key but overlooked factor in the development of biobanks. Genetic epidemiologists are often strong advocates for and important actors in biobank projects, which are portrayed as significant public health initiatives. Genetic epidemiology represents a new kind of surveillance medicine (Armstrong, 1995) that provides a sophisticated and purportedly more objective means than, say, the medical survey used extensively during the twentieth century for monitoring populations according to their degree of deviation from a prescribed ('normal') standard of health. Biobanks provide a vehicle for this surveillance by utilising this knowledge to precisely 'map' categories 'at risk' of disease within the population (the 'genetically susceptible'). The imperatives surrounding genetic health—to undergo pre-natal testing and 'pre-symptomatic' screening for the genetically 'susceptible'—are likely to become pronounced over time, *assuming* that projects endure and useful/useable data are generated. As more and more data are collected, there is likely to be increasing pressure to 'translate' the findings into useful 'outcomes', most likely new genetic tests and drugs that are 'tailored' to certain groups ('personalised' drugs) (see Chapter 2).

It is difficult to see how this translational work can occur without the extensive participation of the commercial sector. The involvement of industry, however, is of concern to those citizen groups that were involved in the consultation workshops organised by UK Biobank during its establishment phase (Petersen, 2005). In particular, fears were expressed that data may be sold to third parties, especially in a context where private profit is likely to prevail over public interest. The establishment of the Iceland Health Sector Database, which involved a single commercial interest, deCODE genetics, revealed considerable public disquiet about this market-driven approach to population-wide genetic research (Rose, 2001a, 2001b). Similar concerns in relation to UK Biobank were voiced by participants in the public workshops organised by the sponsors during the project's establishment phase (Petersen, 2005). In particular, people were concerned that companies would focus on 'profitable diseases' rather than major healthcare issues and wondered whether it would personally benefit them (People Science and Policy Ltd, 2002: 3).

THE NARROW FOCUS OF ETHICAL AND PUBLIC DEBATE

In those countries where biobanks have emerged, the nature and level of ethical and public debate about and critique of their implications has varied considerably. For example, whereas in Iceland and the UK issues of privacy and autonomy have figured prominently in ethical and public debate, in Taiwan these issues have not been evident (see Liu, 2005). Similarly, Biobank Japan, launched in 2003, has thus far attracted little critical attention from lay publics, social scientists, or ethicists (Triendl and Gottweis, 2008: 124–125). In those countries where concerns have been expressed, these have been largely limited to the different expert communities that have a declared interest in this area. The sponsors, custodians, and regulators of biobanks, and the scholars who have researched biobanks, have been preoccupied with a restricted range of issues pertaining to the personal and social implications of such collections. In particular, questions concerning consent, particularly 'informed consent'—its meaning in the context of large, prospective studies of this kind, the most appropriate models of its implementation, and the complexities of its practice—have loomed large in debates at academic conferences and in the professional literature (e.g. Elger, et al., 2008; Tutton and Corrigan, 2004). Debates have, to some extent, also focused on the issue of confidentiality, ownership and commercialisation of data (which, as mentioned, has been the focus of at least some public concern), and 'benefit-sharing'—a loosely defined concept that has attracted growing attention at the international level by bodies such as WHO and UNESCO since the early 1990s (Knoppers, 2005: 11).

The focus on the issue of informed consent is hardly surprising given the ascribed significance of 'autonomy' and 'individual rights' within bioethics and research involving people. In their study of the development, key characteristics, and manifestations of bioethics, Fox and Swazey (2008) note that the concepts of autonomy and individual rights figure prominently in this field, reflecting the values and priorities of the US where bioethics originated. This has influenced responses to normative questions in those countries where bioethics has been developed. The limitations and implications of the use of the concept of 'informed consent' in the context of biobanks are discussed in more detail later in this chapter. It is not surprising that there has also been some concern about ownership and commercialisation, and the distribution of the benefits that may arise, given broader debates within bioethics, the social sciences, feminism, and among lay publics in recent years regarding the privatisation of genetic and health information. Rapid advances in genetic research, combined with growing pressures to 'translate' research into products (e.g. tests, treatments) in an increasingly global market, has heightened fears that that which should be a common resource or a 'global public good', namely, genetic information, may become privately owned and sold to the highest bidder. As Knoppers (2005) observes, increasingly it is acknowledged that traditional legal categories pertaining to property or person are unable to account for the purportedly unique nature of genetic information. Consequently, at the international level, there has been growing recognition and confirmation that the human genome is the 'common heritage of humanity' (Knoppers, 2005: 11).During the early phase of the establishment of biobanks in many countries, discussions about such issues have been largely overshadowed by the predominant concern about the question of how best to ensure that the autonomy and rights of *those who participate* in biobank research are protected. There has been little debate about the value of the biobanks for *the population as a whole or subgroups within the population*, or whether the likely benefits outweigh the financial and other costs, and or whether their governance arrangements are sufficient to ensure that the outcomes feared by critics of biobanks are not realised. The global justice implications of the international development of biobanks, for example, the likely inequalities arising from the resources that they consume, have been largely neglected. Among scientists, policymakers, and those who have come out in support of the development of biobanks, there has been little discussion about the implied commodification of the human body and stored biological material or the reinforcement of biophysical conception of health that such collections entail. The surveillance potential of biobanks, especially with moves to 'harmonise' aspects of different collections and share information, while ever present, has also been largely ignored in the project coordinators' communications, and when concerns have been raised these have been dismissed. Most importantly, no serious effort has been made to develop innovative ways of enabling a broad cross-section of the population to engage in meaningful debates about the substantive issues raised by biobanks. Media coverage of biobank developments has been patchy and mostly lacking detailed coverage of potentially pertinent issues. In those countries where discussions of issues of ethics and governance has occurred, these have, in the main, been framed within a discourse of risk which focuses on either the personal dangers to biobank participants or threats to the research endeavour in the event that a decline of public confidence leads to a loss of funding and political support. Explicitly or implicitly, it is assumed that the application of bioethics principles will ensure the minimisation of potential harms to individuals and the prevention of 'adverse' social responses that may serve to undermine the viability of projects and thwart their presumed future 'payoffs'. The basic scientific premises underlying projects have been mostly unquestioned in the broader public domain.

ASSUMPTIONS ABOUT THE SCIENCE–SOCIETY RELATIONSHIP

The development of biobanks has highlighted some underlying entrenched assumptions about the science–society relationship that shapes science policy in many countries. Science is widely seen to underpin prosperity and community wellbeing and to provide the means of resolving problems that arise from the development and use of technologies. As argued in Chapter 1, however, science is a historical and social product, in that it is sustained through social practices and is heavily shaped by social interests. Further, it always has social impacts and involves 'winners' and 'losers'. The assumption that the benefits from technologies are universally shared is flawed, but is rarely the focus of ethical debates about the impacts of innovations like biobanks. For any technology, there are different constituencies, with pertinent stakeholders differently positioned and affected by the technology. Especially in relation to technologies that involve control over the processes of life, there are many contending positions in relation to their applications and implications. Women and men are likely to be differently affected by such collections, given different gender specific norms regarding responsibility for health and conceptions of a 'normal, healthy' body. Young people's views have also been ignored, which is surprising given that future generations have the greatest stake in the outcomes of any resulting research. Further, as I argue below, those of different ethnicities and religions are known to respond variously to issues such as tissue collection and storage and the 'genetic mapping' of whole populations.

As noted, in a number of countries, concerns have been expressed about various aspects of biobanks, but expression of these concerns alone, with a few exceptions, have generally not led to any shift in the overall aims and direction of those projects. As the events surrounding the Icelandic biobank project revealed, many contextual factors can potentially shape the trajectory of the development of a biobank and that the support of those sections of the population needed for its realisation cannot be taken for granted. Further, they illustrated how the concepts of bioethics—for example, 'consent'—may also be used by those who resist biobanks as well as by those who seek to gain support for them. As Pálsson notes, although it is difficult to draw lessons from this one biobank project, which has some unique aspects, it is interesting to speculate how the project may have evolved with a different set of conditions; for example, the use of informed consent rather than presumed consent, and the adoption of a more collaborative approach between academe, government and industry (2008: 53). Although, as Pálsson observes, there appears to have been a recent shift towards a more 'participatory approach' in bioethics, which is reflected in some national biobank projects (e.g. UK Biobank), in projects developed thus far this has not unsettled dominant conceptions of the relationship between science and society or led to a noticeable shift in power relations. This includes power relations in the economic sphere, which, as Ginsborg notes, is essential if true democratic participation is to take place (2008: 79–91). Rather, as I will show, drawing on data from UK Biobank and some other projects which take its approach as a point of reference, the participatory approach adopted thus far has served as a tool of governance rather than as a means to shape science or to at least generate debate on the relationship between science and society.

THE CASE OF UK BIOBANK

UK Biobank represents an interesting case of how ethics has been utilised in practice in responding to the challenges posed by large scale prospective collections of genetic and personal medical information. While, as mentioned, there is a danger in generalising from the experience of a single biobank project, this project has been taken as a model for the development of other biobank projects in terms of its ethics and governance protocols, and much can be learnt from how it has responded to emergent issues. For example, the Western Australian Genome Health Project, although a very different project in many ways, took many cues from UK Biobank for its ethics and governance protocols and used similar kinds of marketing techniques during its establishment (McNamara and Petersen, 2008). The UK Biobank project is widely cited in the scientific literature and has been the subject of much discussion among scientists, policymakers, philosophers, and social scientists. It developed a 'consultation' phase, with some input from lay publics and healthcare professionals, and it developed an interim advisory group and a subsequent advisory group, whose representation has included ethicists, social scientists, and lawyers. UK Biobank emerged in a context of growing sensitivity about publics' concerns in relation to biomedical practices, especially following a series of medical scandals—including the storage of human body parts (Alder Hey and the Bristol Royal Infirmary cases). The protocols developed for this project can be seen as reflective of a wider institutionalised response to issues involving emergent biomedical technologies. That is, they involve various mechanisms for anticipating, documenting, and containing problems, particularly those pertaining to individual autonomy and wellbeing (i.e. informed consent, confidentiality) (see below). Further, UK Biobank was established in the wake of considerable controversy surrounding Iceland's Health Sector Database, which for many observers was an example of how *not* to develop a biobank.

Given this context and the potential challenges of obtaining and sustaining public support over the expected extensive (decades' long) period of the collection, the supporters and custodians of UK Biobank have made an extensive effort to portray the project as providing potential public benefits, as paying due regard to ethical considerations, and as being competently managed. Despite a number of substantive criticisms of the project by various groups during its establishment phase, including from the UK's House of Commons Select Committee on Science and Technology, the project has proceeded and commenced recruitment of its first participants in 2007. In the nearly ten years between its conception and its launch, the proponents and custodians of UK Biobank struggled not only with technical issues but with its public representations, utilising a number of established bioethics concepts to help establish consent and legitimacy for a project that by its very nature was bound to generate concerns.

Engendering Public and Stakeholder Consent

The architects of UK Biobank seem to have anticipated at the outset some public and stakeholder concerns about large scale collections of this kind, including commercial involvement, protection of personal information, the patenting of data, and the method of recruiting participants. As mentioned, debate about the Icelandic collection, particularly in relation to the commercialisation aspect of the venture and the adoption of 'presumed consent', provided a lesson in how *not* to develop a biobank. Like other recent biobank projects, this one was viewed as a long-term research endeavour with expected but largely unknown clinical benefits in the future. No particular hypothesis was being investigated, and, according to program documents and public statements, the project was never meant to rely on funding from a commercial partner. Concerns about industry involvement were assuaged by the adoption of an 'opt-in' process of recruitment, a method proposed by the pharmaceutical company SmithKline Beecham in written evidence submitted to the House of Lords' enquiry, *Human Genetic Databases: Challenges and Opportunities*, conducted in 2000, around the time UK Biobank was being established.

In its submission, SmithKline Beecham argued that 'concerns expressed about industry involvement in genetic database initiatives can be assuaged by incorporating the best practice developed by companies such as SB [SmithKline Beecham]:

> (i) Protection of patient privacy and confidentiality and using the 'opt-in' approach to participation based on informed consent.
> (ii) Correcting the widespread misunderstanding that raw gene sequence information can be patented and that the patent holder in some way owns that sequence as it exists in individuals'. (House of Lords Select Committee on Science and Technology, 2000)

With a partnership comprising at the outset the UK's Wellcome Trust, the Medical Research Council, the Department of Health, and then later the Scottish Executive, the project has been marketed as a largely public and charity funded initiative rather than a commercial enterprise. This is likely to have added substance to the claim that the aim of the project first and foremost was to undertake research of benefit to the population as a whole.

Being a prospective study, rather than a retrospective study, involving the collection and then storage of genetic and personal medical data over a long period of time, it was crucial that issues of ethics and governance were carefully considered. The project evolved in a context of considerable public resistance to the commodification of the body and its parts associated with biotechnology (e.g. Boyes 1999; Hansen 1999; Pálsson and Hardardóttir 2002), and so it was important that the future *public* benefits of the project were emphasised. In an effort to gain public legitimacy and consent, the project has had to adopt processes that are seen to be transparent and involve the various stakeholder communities and publics from an early stage. Its stakeholder holder-driven approach meant that the method for 'consulting' would always be a limited form of consultation, in that it would address issues seen as pertinent to the stakeholder communities. 'The public consultation' involved workshops comprising small groups of people, many of whom had been involved in previous similar events on different topics. Stakeholder oriented workshops of this kind are mostly limited to involving those who have expressed an interest in the issues, and the potential for expanding decision-making is reliant on prior decisions about their scope (Petersen, 2007a). Rather than taking the opportunity to develop an innovative approach to public engagement, to generate debate about the value of such a collection and about the assumptions that underpin the science, the project's proponents adopted what was arguably a public relations approach oriented to risk management. On its websites and its project documents, they have emphasised its benefits and either failed to acknowledge or downplayed concerns where these have been raised, while underlining that the project will be competently managed and be consultative in orientation. This is not to suggest that the proponents have somehow sought to wilfully 'misrepresent' or otherwise mislead publics to advance the project. Rather, it appears that an early consensus developed among key decision makers that such a project *would* be beneficial—that the economic and health benefits will in time accrue—and that it was therefore worthy of support.

According to GeneWatch UK, in 1999, 'a small group of influential people with close links to biotechnology, venture capital, and pharmaceutical industries . . . began lobbying for a national databases of NHS [National Health Service] electronic medical records linked to individuals' DNA'. It goes on to note,

> The proposal was intended to allow Britain to take the lead in commercialising the human genome and to transform the NHS into a service based on the genetic 'prediction and prevention' of disease, in which large numbers of people could be given 'pre-symptomatic' treatment, massively expanding the drug market. Following lobbying via the House of Lords Science and Technology Committee, this proposal strongly influenced the Government's decision to upload electronic

medical records to a central database (the 'Spine'), at an additional cost of more than £11 billion compared to a localised system. (GeneWatch, 2009: 12)

As the report goes on to note, despite criticisms of this health strategy, the government's belief in its benefits remained unshakeable, and a Ministerial Medical Technology Strategy Group was established in order to drive this agenda forward, 'co-chaired by US company GE Healthcare' (GeneWatch 2009: 12). This occurred without a cost-benefit analysis of the strategy or an assessment of the impact on health or the NHS. UK Biobank was established as 'a pilot project for the planned national genetic database' (GeneWatch 2009: 12). And, notwithstanding scientific criticisms of the design of the project and growing evidence that genes are a poor predictor of common diseases in most people, the Wellcome Trust has forged ahead with plans 'to link DNA databases across Europe in an attempt to make a study big enough to identify very small genetic effects' (GeneWatch 2009: 13). Proponents of 'genetic prediction' and 'prevention', GeneWatch contends, 'have lobbied for researchers—including those from industry—to be able to access information in people's electronic information *without their consent*' (2009: 13; emphases added).

In the program documents and website, the benefits of the project have been repeatedly emphasised:

> Improved means of preventing, screening for and treating these conditions arising from the UK Biobank will have far reaching implications for the health of the public and the health of individuals. . . . (The Wellcome Trust & MRC, 2002)

> . . . Responsibly run projects like UK Biobank are essential if we are to make the best use of the human genome information. They will help ensure that the opportunities for public health provided by these new developments are not squandered. (Professor John Newton, Chief Executive, Press Release, The Wellcome Trust, June 2003)

Program documents have also extensively utilised the language of citizenship, appealing to the concept of altruism and citizens' responsibility to help others (see Petersen, 2005). This suggests that there is a commonality of interest in relation to genetic information, which obscures the substantial concerns voiced by a number of groups (the diverse *publics*) in relation to genetic data. For example, many indigenous groups are concerned that such data may be used as the basis for discrimination or for privatising and commercialising that which is viewed as a community resource. It denies the complex factors underlying the development of diseases and the difficulties of disentangling genetic and other influences on illness (e.g. Holtzman & Marteau, 2000).

UK Biobank was subject to a number of criticisms during its formative phase, including from GeneWatch, which raised concerns about whether the project was 'a good use of public money', and, perhaps most damningly, from the House of Commons Select Committee on Science and Technology which argued that 'a scientific case for Biobank has been put together by the funders to support a politically driven project' (2003: 4). Further, the science itself was criticised by some scientists and members of Parliament, one of whom questioned whether relying on research participants' 'recollections of past behaviour and exposure to environmental risks' may make it difficult to disentangle the genetic and environmental factors that contribute to disease. The fear expressed was that it will 'skew towards over-emphasising the genetic influence on disease processes because it is the only thing on which Biobank will provide hard data' (Gibson, 2002). These were all valid concerns about substantial issues that pertain to many, if not all, biobank projects. However, they are ones that have been largely aired through parliamentary debates and expert forums rather than via more broadly accessible media that would enable lay publics to consider and debate the issues.

DOMINANCE OF ESTABLISHED BIOETHICS PRINCIPLES

When normative questions have been discussed at all within UK Biobank and indeed most other biobank projects, these have tended to be framed narrowly in terms of the challenges posed by such collections for consent, confidentiality, self-determination, and non-discrimination. The development of biobanks has highlighted in particular the limitations of 'informed consent', a concept that was developed originally in the context of single research projects with a limited time frame within Western industrially developed countries. The emphasis on informed consent reflects the emphasis on autonomy and self-determination within bioethics in general. What informed consent means in practice in any context is far from clear (Secko, et al., 2009). Although used since the 1970s, particularly in research involving randomised control trials (Corrigan and Petersen, 2008: 147), the concept is problematic when applied to situations with unspecified future research projects and/or involving diverse communities comprising those with different values, abilities, and circumstances. There is an extensive literature devoted to discussing the issue of informed consent, and particularly its limitations with prospective collections like biobanks, where it is impractical to achieve 're-consent' at a future date when a new research project is undertaken (Fortin and Knoppers, 2009). Some writers have suggested that this requirement limits the usefulness of biobanks and that there should be some loosening of the current consent requirements. Other commentators, however, argue for a more stringent approach to protecting donors' and subjects' interests because of concerns that biobanks may undermine privacy and confidentiality, or be offensive to minority ethnic groups, and enable discrimination (Secko, et al., 2009: 782). As Secko, et al. note, biobanks highlight the tensions between individual interests and broader social impacts and raise the question of what role informed consent should play in biobank governance.

The limitations and adverse implications of 'informed consent' become especially evident when one tries to apply the principle in settings where people are poor, illiterate, or from cultural backgrounds where decision-making by the collective or by community leaders is favoured over decision-making by the individual. In their study of informed consent in genetic research and biobanking in tribal and caste communities in India, Patra and Sleeboom-Faulkner (2009) discovered a number of these and others difficulties of translating informed consent into practice. For example, illiterate individuals were found to be disadvantaged on a number of levels, including understanding the advantages and risks of research and the procedures for fulfilling the formal requirements of informed consent; for example, reading and signing consent forms. Researchers tended to treat illiterate participants in a paternalistic manner. Further, decisions on health and illness were not personal prerogatives, but a matter for family heads or community leaders (Patra and Sleeboom-Faulkner, 2009: 105–106). As the researchers concluded, there is 'a gross mismatch between theory and practical application of the principle of informed consent in settings comprising vulnerable communities in India' (2009: 100).

Questions have been raised about whether people can ever fully understand what they have consented to in the context of large prospective collections with unknown future research deployments. Given the difficulties of re-consent, the custodians of biobanks have often sought to adopt 'blanket consent', which is devoid of information on specified purposes of research and use. It is a way of avoiding the process of re-consent in the future and can avoid logistical and administrative problems (Patra and Sleeboom-Faulkner, 2009: 110). However, this suggests that participants receive adequate information on future potential research options—of which researchers themselves may not even be aware—and calls upon participants to invest trust in the custodians of biobanks to act in line with some conception of the broad public interest. Although endorsed by the World Health Organization in 1997 as likely the most efficient way to use samples in future projects, the concept is especially problematic when used with poor and illiterate populations (Patra and Sleeboom-Faulkner, 2009: 110).

One can question whether consent can ever be 'informed' even with literate populations—if by this it is meant that the individual is fully aware of all potential options and their implications because such awareness is dependent on the context. In theory, options are limitless because they constantly vary through time and across place. Although one may agree to participate in a large prospective research program at one point in time, does this mean that they also agree to participate in related individual research projects that are undertaken at some unspecified time/s in the future? And, is one ever able to foresee the consequences of a decision to participate in research? As Barr expresses it, 'consent is propositional by nature: when one consents to an act, one may be oblivious to the effects of the action one has agreed to' (2006: 260). An assumption underpinning informed consent as it currently operates in practice—that is, as indicated by written confirmation of one's willingness to participate in a research project—is that individual views are unchanging across time and that individuals are fully responsible for the actions based upon those views. However, in reality individual views are very likely to change over time in light of new information, changing personal circumstances, and shifting social relationships. The question of how information is framed and the context in which it is communicated are likely to be highly significant. The context of antenatal care, for example, may shape patients' decisions about whether or not to donate placenta and cord tissue for research. In one DNA banking study, the North Cumbria Community Genetics Project (NCCGP), which involved collaboration between the University of Newcastle Upon Tyne and the Westlakes Research Institute of Northern England, the nature of the sample—umbilical cord tissue that was usually thrown away—appeared to be a significant factor explaining participants' high level of willingness to be involved in the NCCGP. Interviews with donors and biobank managers indicated that afterbirth tissue held little significance for the women, who believed that 'no harm could be done by taking a *non-invasive* sample' (Barr, 2006: 254; emphasis in original).

In light of the above, one can question the value of a form of consent that is focused on individuals as though their decisions are made in isolation of socio-cultural contexts. The limitation of individual consent is especially problematic where genetic information is concerned. In many, if not most, cases involving medical research, others, particularly biologically-related family members, will be affected to some extent in either the short term or long term. The 'subject' of research cannot be strictly limited to the individual who 'consents' but should include all those who are part of the family and/or community. In genetics research, family relationships and whole communities are to some extent affected by studies and are likely to have views on the value and direction of research, a point that was poignantly underlined by the Human Genome Diversity Project (Reardon, 2005). This project, which was proposed by population geneticists and evolutionary biologists in order to resolve fundamental questions about the origins and migrations of species and to challenge the 'Eurocentric bias' of studies of human genetic diversity, highlighted that there are many contending views on the significance of genetics in constructing identity and that conceptions of science and society are always inextricably linked (Reardon, 2005: 1–16). The project also underlined the difficulties of achieving consensus on who should participate in research and the ways in which assumptions, for example, about difference, group histories and lineages, and the potential utility of research findings, may shape discourses of participation.

Within many minority ethnic communities, there is deep suspicion of biotechnology and of biological research that employs racial categories. This suspicion is understandable when one considers the history of expert discourses on race, which have often equated difference with inferiority and have been used to oppress minority ethnic groups (see, e.g. Hannaford, 1996; Reardon, 2005: 17–44). Further, there is deep antipathy to the individualistic approach to consent that underlines most scientific research. This is evident in Māori communities of New Zealand, for example, where the ethical protocol of informed consent, which focuses on a 'single episode' of information giving and assent, sits uncomfortably with the prevailing holistic worldview (Hutchings, 2009). As Hutchings argues, within Māori communities, the notion of collective consent is a more appropriate form of consent than the individualistic form of consent that characterises most biomedical research. Collective consent is congruent with cultural values and collective decision-making and provides the opportunity for members to engage in debate about the significance of new technologies and their impact on Māori cultural values. It also provides a means for collectively resisting health technologies that detrimentally impact on Māori cultural norms and values (Hutchings, 2009: 186–189). In many, if not most, biobank projects, however, the framing of consent in individualistic terms allows little scope for the views of the community to shape research or indeed to decide whether or not a particular project should proceed. If collective consent were to be gained, it is likely that many proposed biobank projects would not proceed or would take a different form from that developed by their proponents. Bioethicists could play a key role in proposing new, collective forms of consent for biobank projects to ensure that deliberation on projects reflects the views of minority ethnic communities and other communities; e.g. disabled people. Their ascribed cultural authority as adjudicators on the rights and wrongs of new biotechnologies and their governance gives them considerable influence in shaping thinking and action. However, this would entail them developing a very different, much broader conception of their role. In the final chapter, I discuss some of the directions in which bioethics might move to make the field more in tune with the challenges posed by emergent technologies such as biobanks.

4   Missing the Big Picture
Bioethics and Stem Cell Research

In February 2010, the following article appeared in *BIONEWS*, a regular newsletter distributed via e-mail by the Progress Educational Trust to those interested in developments in genetics, assisted conception, stem cell research, and related areas:

Is stem cell research being sabotaged by a 'clique'?

> 14 of the world's leading stem cell researchers have expressed concern that truly innovative research may be being suppressed by a small clique of peer reviewers who are intentionally hampering competitor's work from being published in high profile journals. In July last year they sent an open letter to a number of editors of major peer-reviewed journals publishing in the field of stem cell biology. Frustrated by the lack of response two leading stem cell researchers, Professor Robin Lovell-Badge, speaking independently of any institution, and Professor Austin Smith, of the University of Cambridge, decided to make their criticism public by speaking to the BBC last week.
>
> The letter asked the journals to anonymously publish reviewer's comments in the supplementary material online, in order that the peer review process can be made more transparent. Very few scientific journals currently adopt this practice, the European Molecular Biology Organization (EMBO) journal being one of the few exceptions. . . .
>
> (Urner, 2010)

The open letter that is referred to outlined some concerns that the fourteen scientists had, including that 'Papers that are scientifically flawed or comprise only modest technical increments often attract undue profile', whereas 'publication of truly original findings may be delayed or rejected' (Kemp, 2009). The scientists who signed the letter were individuals who worked at prestigious research institutes and universities located in a number of countries.

It would be easy to dismiss these concerns as those of a group of embittered scholars who have failed to produce work of sufficient quality to impact on a field which shows such promise. However, given that it involved some eminent individuals with some distinguished research achievements in the field of stem cell research, the complaints cannot be easily dismissed. The letter, the journal editors' response, and the subsequent scientists' decision to present their case to the BBC highlighted the highly political nature of academic publishing and the issues at stake in the 'race' to make the next 'breakthrough' in this highly competitive field. In particular, it underlined the significance of high-status academic journals in legitimising knowledge and of journal editors in serving as gatekeepers in the struggle for power, money, and status. As the BBC news item highlighted, the consequences of getting published or failing to get published in high-status, 'peer-reviewed' journals are great because the considerable funding in this area (billions of pounds internationally) 'is directed largely towards groups and individuals who have had their research published in the top journals' (Ghosh, 2010). According to the scientists, reviewers who were often competitors in the field were 'sending back negative comments or asking for unnecessary experiments to be carried out for spurious reasons'. They argued that this was being done to stop the publication of the research so that the reviewers or their close colleagues 'can be the first to have their work published' (Ghosh, 2010). Although it was acknowledged that this problem has always existed to some extent, the problem was now more acute, they argued, because research grants and career progression were determined almost entirely by whether a scientist gets published in a major research journal, and the huge amount of funding for stem cell research was providing a greater temptation for those who sought the money to act unscrupulously.

This event not only highlighted the extremely competitive nature of stem cell research and the financial, reputational, and career interests involved, but also scientists' acknowledgement of the significance of the media in shaping the agenda for debate about an issue bearing directly on public knowledge about stem cell science. It also highlights the crucial role of academic publishing in legitimising research which is needed to attract funding and the support of publics and policymakers. The fact that the scientists decided to 'go public' with their grievances reflects their recognition of the potential of the news media to influence public opinion and action. Apart from its airing on BBC News, the issue received extensive coverage in a number of other outlets, including *New Scientist* and the UK quality newspapers, *The Times* and *The Telegraph*. As noted in Chapter 2, many individual scientists and science groups are acutely aware of the role of the media in public debate about science and often actively seek to control the news agenda through the use of public relations, staged news events, and other means. This group of scientists, attached to research organizations of high standing, would have foreseen that breaking their story to a reputable organization such as BBC would attract considerable attention to their issue. In their case, they hoped to influence the reviewing practices of academic journals by having journals publish 'reviews, response to reviews, and associated editorial correspondence . . . provided as Supplementary Information, while preserving anonymity of the referees'. This practice, they noted, had recently been adopted by the *EMBO Journal*, a long-standing and respected journal of molecular biology. Regardless of whether or not their actions had the desired effect, through breaking this story in this way, the scientists involved showed themselves to be shrewd operators with a keen appreciation of their own role in the politics of knowledge. One cannot help but be struck by the contrast between the actions of these science knowledge shapers and the responses of bioethicists who have publically commented on developments in stem cell science over the last decade or so.

As noted in Chapter 1, one of the criticisms levelled at bioethicists is their tendency to react and respond to problems or positions that have been posed by others, particularly scientists and policymakers, rather than adopting an active, agenda setting role, by raising matters not raised elsewhere. In this field of research, perhaps more evidently than others, their preoccupation with responding to a limited array of issues has served to severely restrict debate on a range of major substantive challenges that the field of stem cell research gives rise to. This chapter examines a number of these issues and the implications of bioethics perspectives and concepts as they have been applied in this field thus far. As I argue, stem cell research—both how it is routinely undertaken and how it is represented by scientists, policymakers, regulators, and in various public forums—raises many important normative and justice questions. However, those researching and writing about the bioethical implications of stem cell research have been preoccupied with a limited array of issues defined, in the main, by scientists and/or represented in the media. Their visions and expectations strongly reflect those of the scientists involved in research rather than being based on an appraisal of the likely development of technologies in light of an understanding of the economic, political, and social contexts that shape technologies and responses to them. In their preoccupation with the technologies themselves (rather than the conditions that give rise to and shape them) and with applying abstract principles to discrete issues, particularly the moral status of the embryo and the rights of donors, bioethicists have failed to engage with related global social justice questions. As I will explain later in the chapter, bioethicists could play an active role in shaping science rather than responding to developments to which it gives rise. But to begin developing such a role, they will need a broader conception of their field and a better understanding of the factors that shape both science and bioethics knowledge.

THE PERFORMATIVE CHARACTER OF STEM CELL SCIENCE

Like any field of research, stem cell research is *performative*; that is, it entails the reiteration of certain norms and ways of thinking and acting that enable 'science' to be successfully undertaken, and its outcomes to be 'translated' into useable outcomes; i.e. new treatments. Stem cell research would cease to exist in the absence of a supportive culture, set of institutions and practices, and is thus inescapably a *social* production. An integral, indeed crucial element of this performative aspect of research is the generation of expectations about what treatments will be delivered and when they will be delivered. As noted in Chapter 2, stem cell research typifies the high expectations surrounding biotechnologies. The anticipated regenerative possibilities presented by research in the field of stem cell science at some (generally unspecified) time in the future has led to a considerable expenditure of funds by the public and private sector in many countries. This is estimated to be in the order of many billions of dollars at the international level. The US National Institutes of Health spent over $1 billion alone in 2009, which was nearly double its expenditure of 2006 (National Institutes of Health, 2010). However, among scientists there is considerable uncertainty about the range of treatments that will emerge and the timeline for their availability in the clinic, and about the safety and efficacy of those currently available; for example, as advertised 'direct-to-the consumer' via various private providers (see Chapter 2). Despite a number of clinical trials in process—for example, by Geron, utilising embryonic stem cells for patients with acute spinal cord injuries (approved by the US Food and Drug Administration in January 2009) (http://www.geron.com/grnopc1clearance/grnopc1-pr.html) (accessed 15 February 2010) and ReNeuron's using embryonic stem cells for stroke treatment (Stone, 2010)—as of mid-2010, this field thus far has little to show for this massive investment. Applications are very limited, involving mainly tissue engineered products such as skin grafts or bone repair. Haemopoietic stem cell transplants (using bone marrow, peripheral blood stem cells, or cord blood) for treating haematological (blood-related) conditions such as leukaemias, Hodgkin's disease, immunodeficiences, and inherited metabolic diseases are the most widespread clinical use of stem cells, having been carried out since 1968 (Hollands and McCauley, 2009: 1). Further, by 2010, most research had utilised *adult* stem cells rather than embryonic stem cells. This is not to say that the field will not deliver some or much of what is promised. However, because the anticipated treatments are unlikely to be widely available in the clinic for many years, scientists, the biotechnology industry, and those who support investment in this field have to expend considerable resources in efforts to 'hype' the field.

Like other technologies in their emergent phase, the path of the development of stem cell technologies is uncertain. As with biobanks (Chapter 3) the trajectory of stem cell research and its translation into new treatments is likely to be one fraught with unforeseen hurdles and unintended consequences of actions that are intended to help realise the envisaged positive future of this field. The effort to achieve and maintain funding and public support for stem cell research has entailed a massive international collaborative effort with ongoing actions on a number of fronts, involving science organisations, universities, the biotech and pharmaceutical industries, clinicians, patient groups, politicians, and diverse media, including the print and electronic media. However, these actors do not necessarily hold a shared vision of where the science is heading or of what lines of research should be undertaken. Within science itself there are conflicting views on the value of particular research—for example, that involving human embryos or alternative techniques such as so-called induced pluripotent stem cells (iPSCs)—and on the prospects for the field, the dangers that lie ahead and how best to respond to these. These conflicts are not simply conflicts over the truth of particular ideas. There is a great deal at stake in establishing the public definition of the significance of the field and realising certain imagined futures, including considerable personal and institutional financial rewards, along with status and power. The struggle to control patents—that is, an exclusive right over the technology—is a key issue in this field as it is in other fields of biotechnology research.

The competitive nature of the claims to novelty in stem cell innovations became evident with news reports in February 2010 that Rudolph Jaenisch, a Massachusetts Institute of Technology scientist, 'will be granted a U.S. patent for conceiving a way to turn cells from mammals' bodies into stem cells . . . ' (Waters, 2010). According to one report, Jaenisch claimed to have invented 'the method for reprogramming cells to produce induced pluripotent stem (iPS) cells', for which his company, Fate Therapeutics, had gained the patent. A dispute had arisen between rival parties about 'who filed the patent claim [for the re-programming procedure] first', with Jaenisch obtaining his patent on the grounds that he was the first to make iPS cells 'using genetically-modified skin cells in mice' three years before another researcher, Shinya Yamanaka of Tokyo University, whom 'many consider' 'to be the inventor of this technology' (Waters, 2010). What is interesting about this report is not so much the 'truth' or otherwise of these claims, which cannot be easily determined, but rather its highlighting of the significance of the *representation* of the 'breakthrough' in the 'race' to claim a stake in this field which promises such high rewards. It is difficult for publics to critically assess press coverage of scientific developments such as this, which in many cases originate with public relations arms of research organisations which have a strong vested interest in efforts to promote particular lines of research (e.g. embryonic versus adult stem cell therapies) and in advancing different visions of the field. As noted in Chapter 1, scientists play a key role in science communication by acting as news sources and being strategically positioned to explain the implications of new findings. In this case, the 'President and CEO of Fate Therapeutics' was quoted: 'Dr. Jaenisch's prescient vision in 2003 for creating human iPS [cells], and how reprogrammed cells could be used to revolutionize drug discovery and enable cell-based therapies, is truly unparalleled'. In the same article, another scientist, Jean Loring, 'director of the Center for

Regenerative Medicine at the Scripps Research Institute in La Jolla, California', was also cited, as 'believ[ing] that this may signal a start of a patent war' and as 'dreading the inevitable onslaught of new patents', referring to previous experience showing that owners of such patients have 'interfered with the progress of medical research' (Waters, 2010). Here, readers are presented with two contending portrayals of the significance of this 'patent war': one that foresees new drug treatment of 'revolutionary' proportions and another which portends a future of stifled research opportunities, and hence presumed missed treatment opportunities. Which version prevails will depend crucially on how persuasively and effectively the contending groups can present their claims through media and other forums. In this battle of representations, there is unlikely to emerge a final, generally accepted version of 'truth' about the significance of stem cell science but rather a dominant 'framing' of the field that will, in time, give way to other portrayals, as new competing claims makers colonise the field and succeed in establishing the agenda for debate and policy.

Despite the diversity of views among individual scientists and research teams about the significance and prospects for stem cell research, there is broad consensus within the science community that this field will only survive if it is effectively promoted to influential decision makers (e.g. national policymakers) and 'the public'. Thus, the educational and lobbying activities of science groups such as the International Society of Stem Cell Research (ISSCR) are crucial to the sustainability of the field. The ISSCR aims to 'promote and foster the exchange and dissemination of information and ideas relating to stem cells, to encourage the general field of research involving stem cells and to promote professional and public education in all areas of stem cell research and application' (http://www.isscr.org/mission/index.htm) (accessed 12 February 2010). Towards this end, the ISSCR has sent letters to government leaders (e.g. George Bush) and drafted guidelines for the United Nations and other bodies on issues affecting or potentially affecting research and public support for research, such as proposed legislative changes pertaining to research funding, the use of nomenclature in the field (e.g. 'nuclear transfer' instead of 'therapeutic cloning'), and so-called stem cell tourism (see below). It also hosts an annual meeting involving scientists from public, private, government, and academic organisations. According to its 2010 website, this is designed to 'promote and foster the exchange of research, featuring leading discoveries of the year and serving as a catalyst for inspired stem cell research around the world' (http://www.isscr.org/meetings/) (accessed 2 March 2010). In 2010 it was co-sponsored by the Californian Institute of Regenerative Medicine, Australian Stem Cell Centre, Harvard Stem Cell Institute, along with a range of biotechnology and pharmaceutical companies, including Johnson and Johnson, Geron, StemGent, Pfizer, StemCells, and 'media partners' (Stem Cells and Development, The Scientist, and Wiley-Blackwell) that publish stem cell research and related fields. The site also includes a list of exhibitors from the biotech, medical device, and pharmaceutical industries at the conference, details about 'marketing opportunities' provided by pre-conference advertising, and via the final programs 'distributed with the registration bag', as well as a 'press room' with links for 'ISSCR members', 'scientists', 'public', and 'media'. This sponsorship and this information highlight the close links among science organisations, the biotech and pharmaceutical industries, and the media, including the academic publishing industry, that enable the production of stem cell research. Patient groups are also among the major supporters of stem cell research and often work closely with scientists and health departments in lobbying for research funding in this field and to keep their members informed of research developments, information which is distributed by ISSCR and other science groups. It is with patient activism around stem cell research that one can see clear evidence of an emergent biosociality, a community built upon a shared identity of having a particular biomedically-defined illness or disability (Gibbon and Novas, 2008; Rabinow, 1992; Rose and Novas, 2005). Because the expectations for this field are so high and because there is so much at stake in the outcomes, financially, and in terms of reputations, careers, and people's health, the field has established and maintained considerable momentum since the cloning breakthroughs of 1997–1998 (e.g. 'Dolly') that served as a major catalyst for research (see Thomson, et al., 1998; Wilmut, et al., 1997).

Notwithstanding the momentum of the field, stem cell research has had far from a smooth ride in its development. As noted (Chapter 2), there has been considerable opposition in the US and some other countries about the use (and consequent destruction) of embryos in research. Opposition has originated with diverse constituencies with various concerns, particularly the right-to-life lobby, the Christian Church, and other religious groups (e.g. Islam), who see such research as transgressing religious decrees pertaining to the origins of human life, and with some feminist groups who see such research as exploiting women's bodies and infringing women's rights. In the US, the banning of federal funding of research on stem cell lines created after 9 August 2001 was the source of considerable debate in that country and was the impetus for legislative action in some states. For example, California enacted so-called Proposition 71 (California Stem Cell Research and Cures Act) which enshrined stem cell research as a constitutional right, created the Californian Institute for Regenerative Medicine (CIRM) with an allocated funding of $3 billion over 10 years, and gave priority to research involving embryonic stem cells. Widespread opposition to the use of embryonic stem cells in research in a number of jurisdictions (perhaps most visibly the US) has itself has been productive, leading researchers to explore alternative sources of stem cells which are seen to not carry the same ethical objections, including adult stem cells, umbilical cord cells, iPSCs, and interspecies somatic cell nuclear transfer (SCNT). However, in the US and other jurisdictions, hESC research continues to provide the 'gold standard' for stem cell research because it is seen to offer the best prospects for treatment given that cells can multiply indefinitely and be transformed into other cells (Skene, 2009: 2–3). The election of US President Barack Obama, who removed some of the restrictions on federal funding on medical research involving embryonic stem cells enacted by George Bush, appeared to pave the way for further research with hESCs in the US (Associated Press, 2009), notwithstanding restrictions that apply in respect to National Institutes of Health funding (http://stemcells.nih.gov/policy/2009guidelines.htm) (accessed 15 February 2010). However, while there has been a decline in 'embryo-centric debates' in the US and other countries, the field of stem cell research remains contentious in many jurisdictions, with growing concerns to find ways of better regulating research at the global level through 'harmonizing' (i.e. making uniform) ethical and legal standards (Isas, 2009).

The popularity of the concept of ethical and regulatory 'harmonization' is evident at a time when many different jurisdictions are grappling with the complexities posed by the rapid global diffusion of new biotechnologies. As noted in Chapter 3, a concern with harmonization can also be seen with the development of biobanks. Harmonization suggests that one may apply some universal set of abstract principles or protocols as though it were possible to disregard local histories, cultures, and politics. The underlying assumption is that there exists or could potentially exist a globally agreed notion of human worth and of human rights and set of principles pertaining to human conduct. The tendency to view standpoints and values in a relative perspective, which is a feature of disciplines such as sociology and anthropology, has been strongly resisted in moral philosophy and a number of other disciplines. This is evident in the work of Ruth Macklin (1999), for example, who argues that ethical universals exist and are compatible with a variety of culturally relative interpretations. However, although many values and practices are shared across cultures (and increasingly so, as processes of modernisation and globalisation have taken hold), and notwithstanding wide acceptance of supranational principles such as the United Nation's Universal Declaration of Human Rights, there exists a considerable diversity of concepts of the person, their value, and their rights across nations and cultures. This is clearly evident when one compares concepts of human rights and human worth in, for instance, China with those that pertain in North America, Western Europe, and Australia. However, even *within* societies, different cultural, linguistic, and religious groups may have very different views on issues such as the meaning of life, when life begins, the meaning of 'freedom' and its limits, and so on. As Jane Maienschein's (2003) historical analysis of conceptions of the embryo reveals (see Chapter 2), contemporary debates in the US and a number of other countries about when life begins (at conception, or only later through a process) are not new but have long preoccupied scientists, philosophers, politicians, and lay publics since at least the time of the ancient Greeks. Although science will offer new interpretations of life, science and rational understanding alone will not resolve questions that entail values and are thus subject to contestation and negotiation; that is, that are matters of politics. The question of 'whose view of life?' will prevail (Maienschein, 2003) will depend crucially on who is able to set the agenda for debate and action during the early phase of research and technology development. Despite the frequent claims of many bioethicists to perform a kind of 'horizon-scanning' role in relation to identifying issues and suggesting a path forward on biotechnologies, this is a field where bioethics' commentary and analysis has tended to lag well behind the science. Moreover, the commentary that has been offered has been strongly linked to the concerns of science itself rather than offering a broad assessment of the attendant human rights and justice issues.

BIOETHICS' ENGAGEMENTS WITH STEM CELL SCIENCE THUS FAR

In an effort to gain some insight into the issues dominating contemporary bioethical debates on stem cell science and technology, in February 2010, I undertook a survey of articles published in the area of stem cell research in two major international 'peer-reviewed' bioethics journals, *Bioethics* and *The American Journal of Bioethics*. I acknowledge that these are not the only forums where bioethical debates about stem cell science and technology occur, and it may be that the important discussions are happening elsewhere. However, given that these are key journals in the field and are widely cited by scholars, one would expect that they would reflect the character of bioethics' recent engagements with this field. The aim was not to provide a detailed analysis of the content of individual articles, but rather to identify key themes in the literature and to see to what extent they reflect, reinforce, or challenge key developments in the field of stem cell science and innovation—at least as I understand them. In particular, I was keen to gain insight into the extent to which those who contributed these articles (whom, I acknowledge, may not all call themselves 'bioethicists' or work primarily on issues within the bioethics field) were helping to establish the agenda for public debate and policy. In particular, I was interested to know whether contributors were simply responding to issues raised by others or were helping to set the agenda for debate by highlighting matters not discussed in other forums. A key word search employing 'stem cell' was used, employing the electronic database ISI Web of Knowledge, for each of these journals. I presume this would have picked up any articles where 'stem cell science', 'stem cell research', 'stem cell therapies', 'stem cell treatments', and so on were discussed. This search revealed twenty articles for *Bioethics* and twenty-nine articles for *The American Journal of Bioethics*. After eliminating book reviews and articles where 'stem cell . . . ' was mentioned only in passing, I was left with sixteen articles in *Bioethics*, spanning the period from 2000 to 2009, and twenty articles for *The American*

Journal of Bioethics, covering the period from 2003 to 2009. (Our library holds electronic versions of these journals from the years 1987 and 2001, respectively, and so any articles published beyond these periods should have been captured by the survey.) Given the prominence of stem cell research in popular media and in science during this period, following in the wake of the cloning breakthroughs of 1997, there were not as many articles as I had expected; however, my findings proved interesting.

This survey revealed some recurring themes across both journals. In particular, it highlighted the strong preoccupation with questions concerning *the moral status of the embryo* and *arguments for and against donating or using embryonic stem cells in research and alternatives to using embryos* (e.g. 'Alternative nuclear transfer'). Discussions were often in response to some development emerging out of science, a legislative initiative, or a suggestion on the ethics of stem cell research by some influential body or individual. Much discussion followed in the wake of The President's Council on Bioethics' report, *Human Cloning and Human Dignity* (2002), which revealed a split in members' views on whether embryos have full moral status or none at all and perhaps reflected a division in US citizens' views on this issue (Nelson and Meyer, 2005). For *Bioethics*, 87 percent (fourteen of sixteen) of the articles focussed on issues concerning the moral status of the embryo and alternatives to the use of embryos in research, whereas for *American Journal of Bioethics*, more than 50 percent (eleven of twenty) focussed on these issues. The contributions of some authors (e.g. Baylis, 2008; Lott and Savulescu, 2007) were focal points for response papers on stem cell research related issues; the first on the ethics of using animal eggs for stem cell research, the latter on the advantages of establishing a global human embryonic stem cell bank for overcoming the problem of a shortage of transplantable organs. Illustrative of the concern with the moral questions raised by embryo donation, in one article, it is argued that it is not in the self-interests of female in vitro fertilisation (IVF) patients to donate fresh embryos and identifies some of the 'potential barriers to the autonomous donation of fresh embryos to research' (McCleod and Baylis, 2007). One writer, reflecting on concerns about the destruction of embryos in research and making reference to debates over foetal tissue research and the use of Nazi research data, asks, 'When does benefiting from others' wrongdoing effectively make one a moral accomplice in their evil deeds?' (Green, 2002).

There are a number of nation-specific perspectives on embryonic stem cell research, including from Germany, South Korea, Iran, and Singapore. The German contribution examines 'the logic and morality of the German Stem Cell Act of 2002' which 'permits with qualifications the use of human embryonic stem cell lines created outside Germany before 1 January 2002 (Takala and Häyry, 2007). In another article, a member of Hwang's South Korean research team (see Chapter 2) outlines the donation consent procedures that were developed and used in the derivation of the pluripotent stem cell lines in 2005 (Jung and Hyun, 2006). A short article presents an Islamic view on stem cell research and cloning, explaining that, 'apart from human reproductive cloning, other kinds of stem cell research and cloning are permitted by most Iranian clergymen' (Aramesh and Dabbagh, 2007). Finally, an article examines the 'ethical guiding principles' adopted by Singapore's Bioethics Advisory Committee in relation to research involving adult and embryonic stem cells and cloning, including its support for research involving adult and embryonic stem cells (subject to 'stringent regulations', in the case of the latter) and its objection to the reproductive cloning of humans (Kian and Leng, 2005).

Discussions about the feasibility and ethics of using *alternatives* to human embryos, such as animal embryos, figure prominently in both journals and tend to be in response to recent regulatory initiatives in this area. A decision by the UK's Human Embryo and Fertilisation Authority (HEFA) in September 2007 to authorise protocols allowing human and non-human hybrid embryo stem cell research was the focus of discussion in some articles. A number consider the feasibility and ethics of creating pluripotent stem cell lines by different means, including by 're-programming' somatic nuclei and somatic cell nuclear transfer (SCNT) through inserting human nuclei into both enucleated animal and human eggs. The ethics of so-called cytoplasmic hybrid (cybrid) human embryo research, whereby human DNA is inserted into enucleated animal eggs, widely seen as objectionable on the grounds of it being 'unnatural' or 'playing God', the dangers of crossing species boundaries, and the harm to women egg providers, are discussed. In another article, an author argues against a solution to 'the current impasse over human embryonic stem cell research in the United States', proposed by a member of the US President's Council on Bioethics; namely, using genetic engineering and somatic cell nuclear transfer to create 'pseudo-embryos' that have 'no potential to develop fully into human persons' (Elliott, 2007). Some papers (e.g. Chapman and Hiskes, 2008; Murphy, 2008) respond to a list of ethical objections on cybrid research raised by a particular writer (see Baylis, 2008). Another contribution, in a different vein, proposes that centenarians should be encouraged to donate stem cells, given their rarity and ability to 'resist or overcome the degenerative diseases that kill most of us' (Lewis and Zhdanov, 2009: 1).

These articles address issues of long standing concern in bioethics, namely, the rights, autonomy, integrity, and dignity of the individual, particularly the unborn. They also reflect wider concerns about the manipulation of life and the consequences of transgressing what is seen to be natural that were prominent in the aftermath of the cloning of Dolly the sheep in 1997 (Petersen, 2002). In the wake of this breakthrough, many jurisdictions moved to outlaw human cloning and there was growing concern among many scientists that an 'over-reaction' to this event—especially its 'hyping' or dramatisation in the popular media—would stifle promising new areas of research. The media itself became something of a battleground for competing claims about the significance of the research for human cloning, with many scientists using the media in their role as sources to draw the boundaries between 'legitimate' and 'illegitimate' research—the former defined in terms of that with therapeutic potential and the latter as that which potentially had application in the reproduction of humans. The media also became a forum for debates about the ethical responsibilities of scientists and their 'trustworthiness' in relation to pursuing research for the public interest rather than for professional interest and personal gain (Petersen, 2002: 82–84). The cloning of Dolly sparked many discussions about the ethical, legal, and social implications of human cloning and, in particular, the dangers of altering the natural (see, e.g. Humber and Almeder, 1998; McGee, 1998; Nussbaum and Sunstein, 1998; Silver, 1998) that has arguably carried through into and informed discussion in the bioethics literature as well as policy decisions in subsequent years. That is, the agenda for debate for bioethicists' deliberations on stem cell research has been strongly shaped by the assumptions and expectations of scientists themselves about the prospects and dangers of the field. As is also evident in the field of pharmacogenetics, bioethicists have tended to unquestionably accept—and thereby arguably reinforce—the *technical* expectations that surround this technology (Hedgecoe, 2010: 174). In a sense, bioethicists are, as Hedgcoe so aptly puts it, 'scanning a horizon that has already been mapped out by scientists' (2010: 174).

Of the remaining articles identified, a number focused on issues arising from Lott and Savulescu's (2007) proposal for a 'global human embryonic stem cell bank' to help solve 'an increasingly unbridgeable gap [that] exists between the supply and demand of transplantable organs' and their discussion of the 'immunological challenges' this presents. This includes objections to their proposal to use 'financial incentives to induce minorities to deposit their gametes to a global bank' (Kimmelman, 2007) and to use embryos that (Lott and Savulescu argue) would otherwise be destroyed (Outomuro, 2007), and a piece outlining the relevance of Lott and Savulescu's proposal for the Canadian context where researchers are subject to restrictive policies regarding the derivation of human embryonic stem cells (Green, 2007). One article examined arguments for 'authoritative regulation', particularly via the use of 'procedural frameworks to resolve potential disputes in the public sphere about what is right, wrong, or permissible conduct', with specific reference to the stem cell debate (Capps, 2008). Few articles, however, mapped a new direction in debate about the ethics of stem cell research by raising questions that have been neglected by others; for example, issues concerning research funding and resource allocation in this field and its inequitable impacts, or the underlying views of the body and identity. In *Bioethics*, a notable exception is an article which argued that bioethics has given little attention to the topic of infectious diseases in comparison with other issues such as abortion, euthanasia, cloning, genetics, and stem cell research (Selgelid, 2005). In Selgelid's view, there are a number of reasons for this neglect, including the origins of bioethics itself in its concerns about the dilemmas posed by high-tech medicine, high optimism in medicine which has underpinned the belief that infectious diseases would soon be defeated by medical progress, and the 'religious hijacking' of debates on a range of issues that has served to preoccupy 'liberal-minded bioethicists' (2005: 282–287).

In *The American Journal of Bioethics*, there were also a few articles that placed issues into a broader historical or cross-cultural frame and/or relativised or questioned knowledge claims. An article by Jane Maienschein and her colleagues (2008) examined 'the nature of and obstacles to translational research and assess the ethical and biomedical challenges of embracing a translational ethos' as it pertains to stem cell research (Maienschein's work is referred to earlier in this chapter and in Chapter 2), while another article analysed the arguments against somatic cell nuclear transfer in the Canadian Parliamentary debate to provide insight into the legislative process 'prohibiting a number of research activities, including SCNT' (Caulfield and Bubela, 2007). Such articles were rare, however, and underline the generally constrained character of discussion on this topic and the apparent disinclination of writers to 'push the boundaries' of the field by raising novel questions and by considering the broader economic, political, and social contexts and implications of stem cell research. Further, basic concepts like 'dignity'— particularly in relation to the discussion about embryos (i.e. the 'the inviolable dignity of the embryo')—were employed in a largely uncritical way. (For a discussion on the poor conceptualisation of 'human dignity'—which is much used in bioethics and ethics policies in many jurisdictions—and the dangers of its deployment in policy, particularly the tendency to assume a degree of social consensus that does not exist, see Caulfield and Brownsword, 2006; Caulfield and Chapman, 2005.)

The debates that figure in the international journals, *Bioethics* and *The American Journal of Bioethics*, suggest a disinclination among those within bioethics communities to engage with the 'big picture' issues pertaining to the enterprise of stem cell science and technology, including the rights and wrongs (including the global justice implications) of research funding priorities, the process of knowledge production and 'translation', the impacts of particular research programs and policies, and the personal and social implications of generating unrealistic expectations, including for patients and their families (Petersen, 2009). Hedgcoe's (2010) argument in relation to pharmacogenetics applies equally to the field of stem cell research; namely, that scientists have 'mapped out' the limits to ethical discourse, leaving bioethicists with (apparently) little fertile ground of their own to furrow. As is the case with many, if not most, other fields of deliberation on biotechnology issues, by failing to challenge the assumptions and expectations of science, bioethics arguably serves as a tool for engendering consent and legitimation for the implementation of programs and policies that have been essentially established by others rather than operating as an instigator of debate and action on emergent issues. By reflecting critically on their own constructions of knowledge and reframing their field, however, bioethicists could play an active and influential role in shaping debate and policy agendas in relation to stem cell research. This would entail a substantial reorientation of the field, with a move away from the current emphasis on developing universal abstract principles and greater attention to issues of global politics and power relations, the construction of knowledge, and the particular contexts in which people live their lives. A number of social science projects in progress, including those which have been funded by the UK's Economic and Social Research Council, under its Social Science Stem Cell Initiative, established in 2005, highlight the importance for scholars of adopting a more reflective, critical approach to emergent biotechnologies in the future. Some of this work draws attention to the global politics of stem cell research which bears directly on issues of social justice and highlights implications of and problems arising from the application of bioethics' principles in this field. It not only highlights issues that critical scholarship may take up but also the kinds of approaches that would prove useful in informing debates about the implications of stem cell research. This work emphasises the need for a perspective that pays greater attention to the contexts that shape knowledge and its application.

Global Politics

In their recent book, *The Global Politics of Human Embryonic Stem Cell Science* (2009), Herbert Gottweis, Brian Salter, and Cathy Waldby elucidate the broad global developments under way in the life sciences that have profoundly shaped the expectations for and practices of stem cell science, its 'translation' into useable products, and the national regulatory landscape. These include shifts in healthcare philosophy, which underlie policies of health reform in a number of countries, involving the de-regulation of markets and the changes in thinking and acting associated with neo-liberalism. As these authors note, neo-liberalism includes a number of aspects, including a changed relationship between the individual and the state in matters of health. This is reflected in the adoption of a particular language of health and ethics. 'Self-control', 'self-care', 'consumer empowerment', and individual responsibility for the management of risk increasingly have entered the linguistic repertoire of healthcare programs and policies, with individuals being called upon to take a greater role in managing their own health and wellbeing. Running in tandem with this trend is the process of biomedicalization, whereby 'nature' is harnessed and transformed to create new products and services available in the global, de-regulated marketplace. Regenerative medicine in general and stem cell science in particular has gained widespread appeal in this context, especially given growing concerns in many countries about the health and welfare problems linked to the aging of populations. Stem cell technology offers the prospect of regenerating aging bodies and the problems of degenerative illnesses and disabilities. One of the arguments for the considerable financial support of stem cell science is that the regenerative potential of stem cells will allow for the repair or replacement of organs which 'wear out' as people enter the final years of their lives. Among concerns raised by scientists and bioethicists (see Lott and Sevulescu's (2007) article and responses) is the question of how society will meet the expected growing demand for replacement body parts. A characteristic feature of this global biomedical marketplace is the strong concurrence of interests among key actors in stem cell science: national governments are keen to invest in order to attract capital, gain international science prestige, and contribute to solving problems of health associated in particular with aging populations; biotech and pharmaceutical companies look forward to reaping huge profits from pursuing 'cutting-edge' science and establishing patents; scientists and science research centres expect to prosper through massive government investment; and 'consumers' look forward to technologies that will relieve pain and suffering (Gottweis, et al., 2009: 11–21).

In the view of Gottweis, et al. (2009), ethical reasoning has played an important role in mediating the tensions between the promises of new stem cell technologies and 'cultural costs of scientific advance', and particularly the regulatory challenges posed by human embryonic stem cell research. For the most part, they argue, bioethical actors have supported human embryonic stem cell research and have helped conceptually separate reproductive cloning and therapeutic cloning, which is essential for the effective regulation of hESC research and cloning for medical purposes (Gottweis, et al., 2009: 127). Principlism has served an important role in the 'global moral economy', in that is meshes easily with the requirements of regulatory bureaucracy, providing principles for action and serving to transform transactions between conflicting cultural positions in to apparently value-neutral interactions that can then inform legitimate regulatory policy (Gottweis, et al., 2009: 145). Interestingly, Gottweis, et al. note that the majority of membership of national bioethics bodies are not bioethics philosophers but rather medical scientists and lawyers neither of whom 'has a history of resistance to scientific progress' (2009: 146). While bioethicists have played a political role in mediating the various competing interests in the field of stem cell science, they have had to contend with increasing demands placed on them by the problems of stem cell science, often seeking a response from science itself in search for alternatives to hESCs. The authors argue that while this may lead to some renegotiation of the categories of bioethics with at least some integration of cultural categories of value, they do not envisage other than a temporary pause in the pace of hESC research (2009: 147).

In a separate article, Brian Salter and Charlotte Salter have described in more detail how bioethics has come to operate as a valuable currency—an 'impartial means for achieving fruitful moral trade'—in a global context of many competing views on the status and rights of the human embryo (2007: 559; see also Salter, 2007; Salter and Jones, 2005). In this global context, these authors argue, bioethics' 'success' in encouraging the global moral trade is to be measured by an increase in its own political value and power, particularly if through this trade it advantages scientific progress (Salter and Salter, 2007: 559). Bioethics, in this view, serves a political need, with principlism operating as a tool of governance. Its export to countries outside the US (where it originated) is based on the belief that, like the US, it could help solve problems arising from new health technologies. To achieve global ascendance, it needed to establish ways of marginalizing competitors such as the Catholic Church while demonstrating its unique usefulness to government. It was also necessary that it portrayed itself as inclusive and translate contending moral positions into a common language that was subject to a formal rationality (Salter and Salter, 2007: 561–562). In line with the imperialistic impulse behind the rise of bioethics, over the last decade or so, there has developed an extensive global infrastructure to support its goals. This includes ethical statements launched from established international platforms, an awareness of the need to translate these into a legal/bureaucratic form, and the development of horizontal and vertical networks that link national and international levels of ethical governance (Salter and Salter, 2007: 565). In Salter and Salter's view, the political impetus behind the globalisation of bioethics was the UN Educational, Scientific and Cultural Organization's (UNESCO) *Universal Declaration on the Human Genome and Human Rights*, established in 1997. This Declaration referred to imperatives for human action in relation to the challenges raised by biology and genetics, and it operates alongside the UN body, Intergovernmental Bioethics Committee (created in 2008), which in tandem have help propel further bioethical initiatives by other organizations, including the World Health Organization (WHO) and World Medical Association (WMA). In wake of these developments, it has become unacceptable for any international medical science organisation not to include an ethical element somewhere within its structures (Salter and Salter, 2007: 565).

As Salter and Salter observe, it matters little that members of ethics committees do not include self-identified bioethicists—indeed, for ethics committees of biobanks at least, the majority comprise representatives from medical science (particularly medical genetics) and law (2007: 567). Members' adherence to a shared set of principles and procedures defines their principle identity and constitutes their strength within the 'global moral economy'. In the field of stem cell science, there has been a proliferation of committees that have deemed it appropriate or have been instructed by governments to produce reports on the regulation of research, reflecting the contemporary global political significance of stem cell science (Salter and Salter, 2007: 569). This has given rise to numerous ethical statements and the formulation of 'major ethical trading positions' in relation to the creation and procurement of embryonic stem cells in different countries for which data are available—spanning prohibition to permitting the creation of human embryos for research—presenting a range of policy options (2007: 570–574). Over time, say Salter and Salter, there has been a distinct trend away from a prohibitive stance (represented by the Catholic Church) towards a more facilitative stance in many countries, thus leaving the impression that the Catholic Church is 'fighting a rearguard action that is being gradually lost' (2007: 575). Although those who identify themselves as bioethicists may not dominate bioethics committees, as Salter points out elsewhere, they are emerging as a 'new epistemic power group' involved in 'brokering difficult cultural deals at both national and international levels' (2007: 285). Indeed, Salter boldly predicts that the ascendance of bioethics expertise may over time result in it achieving greater currency than scientific expertise, with bioethicists becoming "the new technocrats" of transnational scientific governance' (2007: 285).

Critical contributions such as these from sociologists and political scientists serve to highlight the significant socio-political role played by bioethics in contemporary societies that are based increasingly on bio-economies. In particular, they highlight the role of bioethics in mediating conflicting value positions and cultural disputes by translating them into matters subject to routine rational management via bureaucratic structures and processes. Such work also underlines the limited potential of bioethics to resolve problems in a field that has built such momentum and involves so many actors with different commitments and investments. By taking the problems posed by science itself as *the* main issues to be tackled, bioethics has failed to offer a critical position on the field as a whole and is poorly positioned to offer direction on emerging developments.

Empirical Ethics

While the above work focusses centrally on the socio-political implications of bioethics discourse and related practices, other recent social science work is concerned more with the disconnection between the abstract principles of bioethics and empirical reality. Recognising the limitations and dangers of bioethics' principlism, a number of sociologists have proposed a more empirically-based analysis of the ethical quandaries posed by stem cell science and other new and emergent fields of bioscience (see Chapter 1). The focus on developing and applying abstract universal principles, many sociologists contend, has led bioethicists to overlook the complexities of socio-political processes and differences in national, cultural, and stakeholder group values and practices. Recent work by the sociologists Clare Williams and Steven Wainwright (2010) illustrates especially well the importance of adopting an empirically-grounded approach to researching the field of stem cell science. In contrast to Gottweis, et al.'s project above, they have focused on the 'micro' dynamics of how those who are engaged in stem cell research and clinicians interpret this field and represent themselves, including their ethical stances. Specifically, they examine the question of how ethical dilemmas and reasoning arise in the clinic and scientists' views on the ethical issues relating to their investigations. Their work draws on in-depth interviews with scientists and clinicians working in the field of stem cell research in clinics in the UK and US, exploring their views on the issues arising from the translation of research from 'bench-to-bedside' in the fields of neuroscience and diabetes. Utilising on concepts from science and technology studies, particularly Gieryn (1999), they explore the 'ethical boundary work' involved in laboratory practice and the 'practical ethics' involved in navigating their path through the various choices as to how they should conduct themselves in a complex political, moral, and scientific context (Williams and Wainwright, 2010). For example, they note that scientists' 'ethical reasoning' justifying the use of embryos in their laboratory work was that these were 'spare embryos' that would otherwise be discarded. As the authors note, such reasoning is also evident in earlier analyses of the arguments of proponents of stem cell research. In their study, some scientists were found to make judgements on the ethics of research according to distinctions based on the *source* of embryos: for 'ethical' reasons, some were not prepared to work on embryos donated from IVF programs but saw no problem with using embryos donated from Pre-implantation Genetic Diagnosis (PGD) (Williams and Wainwright, 2010). As Williams and Wainwright point out, while 'scientists are often portrayed as pushing the ethical limits of biomedical work', those in their sample were in fact *resisting* a more permissive policy on stem cell research as proposed by the UK House of Commons. Their work illustrates the value of 'empirical ethics' in challenging 'normative ethics' and the imposition of principles that are independent of contexts and without reference to the perspectives of those who are grappling with the normative issues that emerge out of research in practice.

Viewed as a whole, the foregoing respective studies emphasise the limits and dangers of bioethics knowledge as currently practiced. The implications of biotechnologies cannot be properly understood without appreciation of their significance within contemporary industrial and industrialising societies increasingly based upon (or expected to be based upon) bio-economies and utilising advanced liberal philosophies and policies. In such societies, science and scientific rationality are ascribed high value, and any contending value systems are likely to be marginalised and to have little influence on the overall direction of policy. In a rational, science-based culture, there is a tendency to instrumentalize knowledge; to translate it into a form that can be easily accommodated with bureaucratic structures and scientific ways of thinking; hence, the value of formal bodies such as ethics committees, advisory boards, and other formal bodies that tend to be dominated by lawyers, scientists, and philosophers (see Gottweis, et al., 2009) whose values and outlooks are broadly in accordance with dominant economic and political ideologies. In this culture, conflicts over questions of value, such as when life begins and ends and the moral status of the embryo, become matters for scientific determination and deliberation according to a certain kind of instrumental rationality.

Some Fundamental Unaddressed Questions

Stem cell science raises fundamental questions about science: about who decides what research should be undertaken, how it should be conducted, who benefits or is likely to benefit from its 'translation' into new treatments (should these eventuate), and the extent to which it is changing conceptions of the body, self, and society. As noted, bioethics tends to uncritically accept the assumptions and expectations of science, particularly in relation to the nature of technologies and how they are likely to be used in practice. However, as noted in Chapter 1, science is an inescapably social production with profound socio-political implications. The expectations of science should be subject to critical scrutiny, with attention paid to the political and justice implications of the visions attached to stem cell science and technologies. Analysis of how these expectations may shape social actions and relationships and limit what is known about the future is urgently required, especially given the massive investments in the field and its impacts on people's lives. Many groups have a stake in the direction and outcomes of stem cell science not just because research in this field involves the destruction of embryos, but because it entails decisions (particularly around funding) which have a material impact on people's lives. As noted, responses, including the development of 'ethics and governance' regulations, are being coordinated and imposed at the global level, with hardly any debate, although they are changing how people interact and view health, illness, and the body.

Any critical perspective on emerging biotechnologies must commence with recognition of the power of knowledge and belief to shape actions. Widespread expectations about stem cell science and what it is likely to deliver in the future are believable and influential because the distinction between science and society has become less and less clear. In science and in popular culture, the question of what constitutes science 'fact' and science 'fiction' is often difficult to discern. Terms taken from science fiction genre like 'designer babies' (e.g. Aldous Huxley's (1955) *Brave New World*) appear regularly in media reporting on the cloning of stem cells, seen clearly in the case of the reporting in the UK of the Hashmi and Whitaker families' plights to use stem cells from 'perfectly-matched' siblings for the treatment of their diseased children (Petersen, et al., 2005). On the other hand, popular cultural metaphors and imagery are pervasive in science, despite scientists' frequent claims that their work is devoid of metaphorical content. Indeed, metaphor is essential to the creativity of science (Maasen and Weingart, 2000). As has become apparent, not only is 'nature' socially produced but it is becoming increasingly *politicised*, in that research in many fields of biomedicine and biotechnology rests upon claims about its alterability and perfectibility, with many debates (e.g. about the moral status of the embryo) being premised upon the veracity of such claims (Petersen, 2007b: 26–30). Competing claims about the pluripotent potential of adult cells and embryonic stem cell research should be seen in this light. Further, evidence on patient group activism highlights how the expectations and representations of science are shaping group identities. The emergence of new forms of sociality ('bio-socialty') and citizenship ('biological citizenship') (e.g. Gibbon and Novas, 2008; Pálsson, 2007; Rabinow, 1992, 1999; Rose, 2007) underline the profound implications of bioscience knowledge for conceptions of the human subject and social action. Finally, new conceptions of the body—as malleable and subject to regeneration—underpin new initiatives in the fields of anti-aging medicine (Petersen and Seear, 2009) and stem cell science (Petersen, 2007b: 39–45). Insofar as bioethics unquestionably reflects sciences' visions, expectations, and arguments, it is incapable of offering a critical voice on the role and potentially far-reaching significance of science in society. As I go on to explain, in the final chapter, a reconfiguring of bioethics presupposes greater reflection upon the guiding assumptions of the field and a better appreciation of how certain long-established antecedents continue to constrain thinking and actions. The next chapter, which focuses on genetic testing and counselling, provides a case in point, with deeply entrenched assumptions about human subjectivity, the significance of knowledge in human action, and the role of expertise having a firm hold over current policies and practices in these areas.

5   Testing Times
Bioethics and 'Do-It-Yourself' Genetics

Genetic testing to get flawless babies

> Couples are to be offered a groundbreaking genetic test that would
> virtually eliminate their chances of having a baby with any one of more
> than 100 inherited diseases. The simple saliva test, which identifies
> whether prospective parents carry genetic mutations that could cause
> life-threatening disorders such as cystic fibrosis, spinal muscular atro-
> phy or sickle-cell anaemia in their children, is to be launched within
> weeks in Britain. . . .
>
> (Henderson, 2010: 3)

Special offer for 'genome scan' customers willing to participate in research

> Personal 'genome scans' have been in the news again, not because they
> are a particularly new phenomenon but because the data they provide
> may soon take on a new significance. Up until now, genetic tests look-
> ing for variations associated with diseases have been available over the
> Internet for those who have been willing to pay the fee—usually a few
> hundred dollars. Now US company 23andMe has just launched a new
> home DNA test for $99 that is being marketed as part of the 'Research
> Revolution'. . . .
>
> (Beauchamp, 2009)

The age of direct-to-consumer advertising (DTCA) of genetic tests has
arrived. The first of these articles refers to the marketing of a test via Bridge
fertility clinic in London (where it is offered in conjunction with genetic
counselling) and 'directly to customers' from Counsyl, 'the US company
that developed the test'. As the article notes, the test will be offered 'directly
to customers over the Internet, for home use without medical advice' (Henderson, 2010: 3). An examination of Counsyl's website, indeed, reveals the
ready availability of their genetic tests via the internet, or the '100+ clinics
across the US'. In this case, the marketing is clearly targeted at prospective parents who are informed that the test can be easily ordered online,
undertaken in minutes, with results available 'within 2 to 3 weeks' (https://
www.counsyl.com/) (accessed 26 March 2010). Visitors to the website are

advised that such tests are a means to 'Protect your child from 100+ genetic diseases', and that 'unsuspecting couples are at risk of conceiving a child with a serious genetic disease, such as cystic fibrosis, spinal muscular atrophy, or Tay-Sach's disease'.

The second article is one of a number that have appeared in recent years announcing the availability of 'personal sequencing' services via the internet. In this case, however, unlike other such tests, this one 'is only available to participants who allow both the genetic data collected and the health data they report to be used by researchers investigating the occurrence of different diseases'. It also reports that this is not the first company to use this approach to DNA testing, with another company, TruGenetics, a Seattle-based company, launching such a test with the first 10,000 participants signing up getting DNA tests free of charge (Beauchamp, 2010). A visit to the websites of 23andMe and TruGenetics reveals that they offer similar kinds of services; namely, the offering of a 'simple' genetic test via the use of a sample of saliva to create a 'personalized health record', followed by the relatively quick feedback of results. Like Counsyl, 23andMe and TruGenetics heavily utilise the language of personal empowerment: for example, 'empowering everyone to explore their genome and advance scientific research'; 'TruGenetics™ will empower every individual, regardless of race or socioeconomic background, to explore their genome' (TruGenetics—http://www.trugenetics.com/) (accessed 26 March 2010); 'Choose the DNA test that is right for you'; 'Take charge of your health'; 'Choose to have it all'; 'Let your DNA help you plan for the important things in life. Take charge of your health and wellness today' (23andMe—https://www.23andme.com/) (accessed 26 March 2010).

The development of the internet-based advertising of genetic tests, like other forms of DTCA (e.g. pharmaceutical products, and stem cell treatments) (see Chapter 2) is a manifestation of the profound changes that are occurring in the practices of health and healthcare. These changes are not limited to the established institutions of healthcare, but are part of broader re-shaping of identities and relations within and across societies. The fact that some of the above companies have sought to entice 'consumers' to undertake tests in order to collect genetic tests for research—a kind of bio-banking *without* consent (see Chapter 3)—is an innovation with potentially wide-ranging impacts that has been neglected in the academic literature and is difficult, if not impossible, to regulate. This is an area where technological developments are running way ahead of the ethical and regulatory responses. DTCA in general illustrates the inseparability of science and society and how the growing convergence of technologies (digital and genetic) is generating new possibilities for action and new conceptions of self and society. DTCA of genetic testing specifically reflects the intensification of the commodification of life and the de-regulation of healthcare that is occurring internationally. The advancement and convergence of new technologies—particularly genetic and digital technologies—has created the potential for new tools for manipulating and marketing data and for monitoring individuals and populations. Discussions about the implications of genetic testing thus far, however, have been limited by an ethics discourse focusing on a limited array of issues, such as how best to ensure the achievement of 'informed consent', how best to guard against 'coercion' in genetic decision-making, how to balance the 'right to know' with people's 'right not to know' about their genetic risk, and how to ensure 'non-directive' counselling in the case of those who are offered information that is derived from genetic tests. Concern about 'autonomy' and its preservation underlines many discussions. Through drawing attention to this limited array of issues and its utilisation of the language of individual rights, choice, and empowerment, bioethics knowledge has served to obscure the far-reaching politico-economic and socio-cultural implications of the 'geneticisation' of life.

This chapter examines the role played by bioethics knowledge in a global healthcare marketplace of genetic tests that are increasingly marketed directly to consumers. In the 'post-genomic' era, gene testing has emerged as an extensive enterprise within healthcare, contributed to and supported by many health professions, patient organisations, research organisations, and ethical and regulatory bodies. More and more, genetic tests have been 'mainstreamed' into healthcare in many countries as the 'genetic worldview' (Miringoff, 1991) has taken hold. Genetic tests provide a means of classifying conditions and 'sorting' populations: 'diagnostic' tests (to confirm or disconfirm a genetic condition, for those with symptoms); 'predictive' or 'pre-symptomatic' tests (for healthy, high-risk individuals who may develop 'late-onset' monogenetic disorders); 'pharmacogenetic' tests (to determine susceptibility to adverse drug reactions or the efficacy of drug treatment in those of a particular genotype); 'carrier' tests (to detect a mutation that has limited or no consequence for the individual but may confer a high risk in offspring); 'prenatal' tests (performed during a pregnancy to detect a heightened risk of a condition in the foetus); and 'pre-implantation genetic diagnosis' (PGD) tests (undertaken to detect the presence of a mutation, in order to select the unaffected embryos to be implanted, in the case of assisted reproduction). (EuroGentest—http://www.eurogentest.org/professionals/info/public/unit3/final_recommendations_genetic_counselling.xhtml) (accessed 2 April 2010).

As this list reveals, genetic tests are seen to have wide application in healthcare as a tool for the management of risk and personal decision-making in health and beyond. A culture of testing is becoming entrenched in healthcare, with patients as well as healthcare professionals looking to utilise such tests to diagnose what are often problems of a complex aetiology. Increasingly, they are viewed as a means to enhance one's ability to plan for life, including for reproductive decisions, career choices, and decisions about intimate relations (e.g. 'genetic matching', in some communities). Their burgeoning use is congruent with the increasing emphasis on 'care of the self' evident in many spheres of social life (see Chapters 1, 3, and 4). With the growth of genetics knowledge, the distinction between testing to learn about health status and testing in order to inform decisions about lifestyle and enhancement has become blurred. Testing is portrayed as offering the opportunity to eliminate disease and 'design' the 'perfect baby' and as facilitating 'choice' in healthcare decision-making, thus serving as a tool of 'empowerment'. There are many implications arising from the proliferation of genetic testing in healthcare, for how we view the body, health and illness, self-identity, and relations to others within families and communities. The potential for discrimination on the basis of genetic difference and for the systematic termination of 'imperfect' foetuses is significant and has been commented on by many writers who are concerned about the implications of such testing. However, within the bioethics literature and among those who deliberate on the 'rights and wrongs' of genetic testing, the predominant focus has been on efforts to protect the autonomy of the individual who is undergoing testing, particularly through adherence to practices of 'informed consent'.

Building on a number of points raised earlier (Chapters 1 and 3), the chapter highlights the implications of this focus on the concepts of autonomy and informed consent as it pertains to the practices of genetic testing. As noted, the concept of informed consent is premised upon a construction of the human subject as an autonomous rational decision-maker whose 'choices' are unconstrained. In this idealised view of human subjectivity and decision-making, freedom from coercion or constraint is advanced by more or 'better' information 'packaged' in an appropriate form. There are many problems with this conception of the human subject, reasoning, and action, some of which have been touched on in previous chapters. This includes the tendency to ignore the process by which information is constructed and mediated and the power relations that shape interactions and the framing of information and its reception. A major criticism of the conception of 'informed consent' in bioethics is its individualistic emphasis. As a number of writers have argued, the effects of genetic information are not limited to the individual who is tested; by its nature, it has consequences for others in one's family and communities (e.g. Hallowell, 2009; Rehmann-Sutter, 2009). Those who undergo genetic testing have responsibilities to others, a fact that is recognised by many individuals themselves who report that their decisions are influenced by their perceptions of other family members' needs (e.g. Hallowell, 1999, 2009). As Hallowell, et al. argue, in their study of women who had been diagnosed with breast or ovarian cancer and subsequently tested for BRCA1/2 mutations, the concept of informed consent fails to acknowledge that those who undergo genetic testing (in this case, women) are 'relational entities in both the biological and social sense', and that individual 'choices are constrained by virtue of the fact that they exist within a network of relationships' (2003: 78). Elsewhere, Hallowell (1999) has shown that women perceive themselves as having a responsibility to their kin to establish the magnitude of their risk and risks to other family members. However, in seeking to 'do the right thing', they 'relinquished their right not to know about their risks' and undertaking risk management practices that posed their own dangers. As this work makes clear, 'autonomy' and 'empowerment' in the context of genetic testing are far from unproblematic concepts. However, they continue to strongly inform bioethical reasoning and policies in genetics-based healthcare, including those pertaining to genetic counselling, whose premises and implications will be critically scrutinised later in the chapter.

As noted in Chapter 3 in relation to biobanks, the limits of informed consent have become especially evident as genetic, personal medical, and lifestyle information is collected over the longer term and the nature of the research based upon that information is unspecified at the outset. Whereas individual consent may have some utility in certain circumstances (for example, in projects or interventions with short time frames and where the research purposes are known and clearly specified at the outset) (see Chapter 1), a preoccupation with informed consent—as though it were the most important or only issue to be resolved where technologies are used—diverts attention from the profound social implications arising from the development and utilisation of new biotechnologies. As in other areas of biotechnology discussed in earlier chapters, in the field of genetic testing, bioethics concepts serve a legitimising role. By showing that the requirements for informed consent have been adequately attended to in research and clinical practice (through a kind of 'tick-box' process), bioethics bolsters the power of science and deflects attention from substantive questions pertaining to the direction and consequences of the development of particular forms of scientific knowledge and related technologies.

INFORMED CONSENT AND NEO-LIBERALISM

A fundamental problem with clinically-based informed consent as it has been conceptualised within bioethics is that it has emerged within and been shaped by a set of institutional arrangements and practices that have undergone or are undergoing rapid change under the influence of neo-liberal philosophy and policy. In particular, it overlooks the reconfiguration of the relationship between citizens and the state and changes in supportive institutions and policies that have occurred in many countries since the mid-1970s (see Chapter 1). This includes a greater emphasis on self-governance (e.g. care of the self, individual management of risk), the 'downsizing' of government with a greater reliance on the private sector for the provision of a range of basic services, especially healthcare, and the substantial liberalisation of markets assisted by various policies (e.g. taxation) and programs (e.g. social insurance). One of the significant changes that has occurred during this period has been the rise of the internet and ubiquitous DTCA of healthcare products and services and their easy purchase online and via practitioners that are not medically trained (e.g. pharmacists, alternative health practitioners). I referred to the significance of the DTCA of stem cell treatments in Chapter 2. A range of medical treatments and tests are now marketed by DTCA, including pharmaceutic products (Donohue, et al. 2007), bariatric surgery for obesity (Salant and Santry, 2006), and genetic tests for hereditary ovarian and breast cancer, predicting gender, pregnancy risk, risk of thrombosis, and other conditions (Goddard, et al. 2009; Henen, et al., 2010; Stein, 2010; Williams-Jones, 2006). A great deal of healthcare now occurs outside the clinic, with self-testing, self-care treatment and services, and self-medication (including use of complementary and alternative medicines) constituting a huge and rapidly growing global business.

There are now a large number of websites devoted to patient self-care, offering information on treatments and the state of research on particular conditions, patient narratives, and so on. For example, the website Patientslikeme.com in mid-2010 included patient profiles on approximately 60,000 members, with site users being able to filter information according to type of treatment, symptoms, and patient characteristics. With information on research tools, clinical trials and means for assessing the status of one's condition 'relative to the rest of the community'—presented as a collaborative endeavour involving patient groups, academics, and industry leaders—the site presents as a kind of benevolent, altruistic enterprise devoted to those seeking to manage their own condition. One does not have to search too far on this site, however, to see that this is no 'purely altruistic' enterprise with pharmaceutical giants such as Eli Lilly and Company, and Novartis mastheads prominently displayed alongside the descriptions of some research initiatives (see http://www.patientslikeme.com/all/patients) (accessed 15 June 2010). Internet-based patient 'resources' such as this, along with other forms of DTCA via pharmacies, shopping centres, and other outlets, is rapidly changing the face of healthcare, with evidence of commercial influence apparent at virtually every level. The underlying premise of the concept of informed consent, however, is that healthcare and research mostly occurs in institutional settings characterised by particular expertise and relationships, and certain checks and controls on commercialisation and exploitation.

Within the literature on informed consent, the process of consent is conceived as adherence to a set of principles; namely, that relevant information has been given, that information has been understood, that the person is capable of offering consent, and that the person is not coerced (Holm and Madsen, 2009: 12). This presupposes the existence of certain roles (importantly, the information provider, who is assumed to possess and present the requisite 'package' of information in an appropriate form; and the research or clinical subject, conceived as a competent, autonomous rational information processor), relationships (the expert and the lay subject or patient), and settings (typically, the clinic). Within the healthcare setting, an important 'enabling'/facilitating role has been ascribed the genetic counsellor, who through the adoption of a 'non-directive' approach ensures that the patient/'client' arrives at their decisions 'autonomously'—'empowerment' being the aim. As Rehmann-Sutter argues, the historical significance of the principle of non-directiveness has been as a defensive tool in the formation of medical genetics by 'upholding the idea that the medical offer of prenatal testing is morally defensible and innocent of eugenics' (2009: 118). As I explain later in the chapter, eugenic outcomes may arise, however, regardless of intentions, as a consequence of the focus on individual empowerment. The process of achieving consent has become established over time as a largely routine (i.e. institutionalised) affair; that is, it entails adherence to an agreed set of protocols and rituals (e.g. adherence to research ethics guidelines, the submission of an ethics application to an ethics committee constituting appropriate expertise, the presentation of consent forms and information sheets to participants, the recitation of the 'subject's' right to withdraw at any stage of the research, and the 'subject' 'giving their consent' via the signing of the consent form). In other words, these social arrangements have fabricated and addressed human subjects in quite specific, standardised ways mostly without reference to the socio-political context shaping thinking and action. While this may serve to fulfil established ethico-legal requirements—with due process being seen to have been rigorously followed—it fails to acknowledge the complexity of individual decision-making and the longer-term social consequences of individual decisions. As many researchers and practitioners from diverse disciplines and areas of practice have come to recognise, the drive to standardise the process of consent has led to its reification. The routine, reified nature of the process of consent is reflected in the use of particular language—for example, reference to the 'consenting' of 'subjects'. Recognising the limits of informed consent, some writers (e.g. those contributing to Corrigan, et al.'s (2009) recent collection) have argued for an additional means of protecting the rights and welfare of patients and healthy volunteers who are involved in research.

Authorities' Response to DTCA of Genetic Tests

The marketing of genetic tests directly to consumers via the internet and other avenues (e.g. via pharmacists, alternative health practitioners) has been of some policy interest in recent years, with authorities seeking to regulate the field through codes of practice and public educational initiatives. For example, in 2003, the Human Genetics Commission (HGC), which advises the UK government, published a report, *Genes Direct: Ensuring the Effective Oversight of Genetic Tests Supplied Directly to the Public*, which offered a series of recommendations to regulate what it saw as a number of concerns in this area; namely, 'the danger that the public might receive misleading medical advice as a result of companies overstating the role of genetics in common complex diseases; the difficulty of ensuring informed consent when tests are offered direct to the public; and the impact on NHS resources if patients were to seek advice from their doctors before or after tests, or if patients were to require confirmatory testing within the NHS' (citing Human Genetics Commission, 2007: 5). This report focused on *predictive* genetic tests and recommended that tests should *not* be marketed directly to the public but indicated that where they were they should be subject to a code of practice. It emphasised the need for the pre-market review of tests to assess 'the scientific and clinical validity and clinical utility of genetic services and whether they ought to be offered directly to the public' (2007: 5). The report also recommended funding for educational initiatives 'to improve public understanding of genetics and the development of impartial sources of information on genetic tests' (2007: 5). In other words, reflecting the bioethics focus on 'autonomy', the emphasis in the report was on ways to promote a more 'informed' decision-making by 'consumers' through regulatory and educational efforts. The report offered no critical comments on the underlying consumerist approach to healthcare, the implied commodification of the body and health, the potential inequitable and discriminatory impacts of the widespread marketing of such tests, and the considerable challenges of regulating genetic tests in an increasingly de-regulated healthcare arena.

At the time of the publication of this report, few genetic tests were marketed directly to consumers via the internet; however, anticipating (correctly) that there was likely to be a growing number of genetic tests on offer for a range of conditions, it further recommended that the take up and continuing relevance of the recommendations be reviewed in three years. By November 2007, as noted in the Appendix of the follow-up report, *More Genes Direct: A Report on Developments in the Availability, Marketing and Regulation of Genetic Tests Supplied Directly to the Public*, genetic tests were available for various purposes. These included the assessment of 'genetic health' (including predisposition to breast cancer, prostate cancer, osteoporosis, thrombosis, and cancer), 'predisposition to nicotine addiction and response to nicotine replacement products', 'fetal DNA gender', 'alcohol metabolism', 'asthma drug response', 'athletic performance', 'Alzheimer'', 'pregnancy risk', and 'drug sensitivities' (Human Genetics Commission, 2007: 31–36).

Some of these 'conditions' (e.g. athletic performance and addiction) are not of a *medical* nature but rather are phenomena or traits that are subject to strong social influence. 'Fetal DNA gender' presumably refers to sex selection, which is a contentious process in many societies because of its discriminatory implications, although increasingly used in prenatal decisions by some parents in some countries. Gender, of course, is not a 'medical condition' but a socially constructed phenomenon, with current thinking in gender studies being that it is defined by its performative character; by certain reiterative practices (in other words, we *do* or *enact* gender rather than *have* a gender) (West and Zimmerman, 1991). In this case, it is interesting to note how a biomedical technology is being marketed to assist family planning decisions that are personal matters that are socially shaped. The increasing availability of tests for such phenomena or traits indicates the 'geneticisation' of life, referred to above, where virtually all aspects of life have been explained in terms of genetic traits, 'defects', or 'faults'. This process is very likely to become more intensive as information derived from genetics research is translated into new genetic tests in the future. Interestingly, the report has nothing to say about the regulation of genetic testing for these non-genetic or only partly-genetic phenomena, which may involve the exploitation of consumers while reinforcing the genetic worldview.

Belief in the power of genetic information is contributing to the blurring of the boundary between interventions for treatment and interventions for enhancement. While in healthcare, genetic information is portrayed as a means to improve health decisions and outcomes in the wider culture, it is also seen as offering an explanation of human behaviours and attributes; for example, as explaining differences between men and women and between people of different sexual orientation. This belief in the power of genetics is reflected in science as well as popular cultural media (e.g. Petersen, 1999). Among the recommendations in the above 2007 report were that the marketing of tests that 'should only be available via a suitably qualified health professional should be restricted (i.e. no direct-to-public advertising'), and that 'The use of existing web-based information sources to provide comprehensive and independent information for consumers should be explored' (Human Genetics Commission, 2007: 24–25). That is, the recommendations reinstate a conventional view of the role of experts (as dispensers of information) and of the factors that enhance patient autonomy (i.e. the provision of 'appropriate' expert knowledge), noted above. Confirming the conclusions of the 2003 report, in summary it was noted that genetic tests that are available over the internet cannot be easily controlled and that the best that could be achieved was to promote 'high standards of regulation in the UK' and to liaise with regulators in other countries in order to 'achieve effective and harmonised national and international controls' (2007: 25). Further, while it acknowledged the difficulty of providing up-to-date information on every genetic test on a voluntary basis (2007: 19), the report did not question the assumption about the value of such web-based information or the role of test developers/providers in facilitating consumer access to this information. In short, the message conveyed is that this is an area which is subject to limited effective regulation by authorities and would need to be heavily reliant on self-regulation and the good will of test providers, with patients themselves being encouraged to play an important role in informing themselves about the utility of genetic tests through being better educated and gaining access to 'independent sources of information' (2007: 19). As this book was being completed, the UK's Human Genetics Commission (HGC) published *A Common Framework of Principles for Direct-to-Consumer Genetic Testing Services* (2010), which essentially continues the focus on the provision of information to prospective consumers, along with an appeal to test providers to act responsibly in advertising tests and complying with any legislation or voluntary codes. Again, there is a call for counselling (pre- and post-test) and informed consent, and a demand that test providers seek to ensure the security and confidentiality of information and maintain high standards in professional practice (Human Genetics Commission, 2010: 6–13). In short, the principles articulated in this report are not essentially different from the recommendations proposed in the earlier HGC reports, the focus being firmly on the means to protect the autonomy and rights of prospective consumers.

Some authorities acknowledge the problem of heightening expectations of the value of genetic tests for individuals when these will be mostly limited to single gene disorders, rather than conditions which arise from the interaction of multiple genes and genes and environments. An example is the UK report, *Genomic Medicine: An Independent Response to the House of Lords Science and Technology Committee Report*, published in May 2010. In brief, this acknowledges the failures of the House of Lords report—which focused on the current and future impact of genetic technologies on clinical practice—stating that this report,

> failed [to] adequately reflect the realities of genomics and health, placing too much emphasis on the prospects for predicting the risk of common diseases for individuals, whilst paying too little attention to the current opportunities to use existing and rapidly emerging genetic knowledge and technologies to improve services for the diagnosis and care for single gene disorders. Although genomics has increasing potential to offer better care for patients outside the traditional sphere of clinical genetics—including through the use of pharmacogenetic tests to improve the efficacy and safety of drug treatments and genetic tests to help identify rarer inherited forms of common diseases such as cancer—risk prediction based on genome profiling has very little utility for individuals. (Brice, 2010)

However, despite occasional acknowledgement of the limitations of genetic testing for the prediction of common diseases, official responses to the increasing availability of genetic tests have not sought to challenge the discourses of autonomy, individual rights, and choice. For example, there has been no effort to generate a debate about the underlying consumerist approach to healthcare, the feasibility of efforts to control or ban 'direct-to-consumer' genetic testing, or the value of research into developing the technologies themselves. As in many other areas of biotechnology, the technologies themselves have been taken as given or 'pre-social', rather than as reflecting social priorities, social relations, and visions of how society should be organised. The philosophy of healthcare informing the adoption of genetic tests in healthcare decision-making remains mostly unquestioned.

The Significance of 'Choice'

'Informed choice' is a key guiding principle in contemporary neo-liberal healthcare and has provided one of the key arguments for increased individual access to genetic information in recent years in many countries. As I have argued previously (Petersen, 2006; Petersen and Bunton, 2002), according to the discourse of the new genetics, more or 'better' information is assumed to provide the foundation for more choice which allows greater freedom in healthcare decision-making. Along with the assumed benefits that will accrue for 'the public's' health, enhanced choice has been a key rationale for 'mainstreaming' genetics into healthcare in recent years. This assumption is shared by various practitioner communities, biotechnology and pharmaceutical companies, and venture capitalists who seek to exploit the opportunities presented by the 'genetics revolution'. DTCA of medical products has been exploited by those marketing gene tests as a means to 'empower' patients. However, as Williams-Jones (2006) notes, drawing on the examples of such advertising for pharmaceutical products and genetic testing for BRCA1 and BRCA 2, such advertising has not undermined the professional power of healthcare professionals. Rather, patients are encouraged to actively participate in healthcare decision-making by seeking appropriate professional advice and increasing their awareness of treatment options. Such advertising seeks to induce patient demand for brand name drugs and tests by utilising a variety of media, particularly the internet, 'exploiting a climate of genetic determinism and the public's misunderstandings of and anxieties about susceptibility, probability and risk' (Williams-Jones, 2006: 95). It supports and reinforces a new form of citizenship, 'biological citizenship' (Rose and Novas, 2005), whereby individuals take an active role in managing their own condition by becoming informed, active consumers of health information and new treatments. It is an integral element of the 'political economy of hope', whereby patient expectations for positive outcomes of treatments are exploited in marketing products which promise more than they can deliver (see Chapter 2). The internet is a key communicative tool, enabling advertisers to provide potential consumers with a wealth of information that, given constraints of costs and time, would otherwise be unavailable through other media (Williams-Jones, 2006: 96).

Given the focus on 'consumer choice', it is hardly surprising that the *More Genes Direct* report should emphasise the importance of the provision of information from 'independent sources' to counteract the perceived risks arising from the advertising of genetic tests directly to consumers

(Human Genetics Commission, 2007: 19) In this report, it is suggested that the Human Variome Project, 'an international initiative to link all genetic databases', and Lab Tests Online, advertised as an 'online source of information designed to help patients and carers understand better the laboratory tests used in the diagnosis and treatment of disease', are promoted as potentially two useful sources for informing consumers about new tests as they become available. It is worth examining these sources in some detail because they reveal the considerable momentum that has been established in making genetic information widely accessible to the population as well as the network of actors that have collaborated in the effort to realise the envisaged genetics-based healthcare future. Such resources are likely to gain increasing significance as repositories of data for 'informed consumers' as well as researchers and policymakers in the future once their supportive physical and regulatory infrastructure (with support by ethical protocols) is established and information is globally shared. Although these particular sources do not offer genetic tests directly to consumers, the detailed information that they provide on genetics and genetic tests lends the impression that the value of such tests has been established. They are thus potentially influential sources for those seeking information about them. The endorsement of the websites by various health organisations also implies that tests have been widely authorised in professional communities and are thus legitimate, despite frequent acknowledgement of the problems that accompany their interpretation. One has no way of accurately assessing how widely such sites are used by those in search of medical treatments and the extent to which they shape individual actions. However, growing evidence of the influence of DTCA on healthcare decision-making (e.g. Australian Medical Association, 2008; Mansfield, et al., 2005) and of a high 'hit rate' on some sites (e.g. the UK's Lab Tests Online had 100,000 visitors in January 2006) (Association for Clinical Biochemistry, 2006: 3) suggests that such sources are likely to be influential among patients and/or their families who are making healthcare decisions.

The Human Variome Project: The Creation of a Global Databank

According to an Editorial article in *Nature Genetics*, that is linked to its website, the Human Variome Project (HVP) is a 'successor to the Human Genome Project' that aims to develop systems to collect and disseminate data on genetic variations associated with human disease generated by various clinical and diagnostic laboratories throughout the world. This will create a comprehensive resource for researchers and 'add value' to existing genetic databases such as HapMap, GenBank, and OMIM (Online Mendelian Inheritance in Man). The HVP is making efforts to standardise classifications and collection systems at the global level and to upload information to the web to provide an accessible 'encyclopaedia of data' for researchers, clinicians, counsellors, support groups, patients, and 'instrument and supply companies'. As part of its repertoire of activities, this project involves country-specific meetings; pre-meeting forums 'to promote discussion and generate new approaches and collaborations'; working group meetings 'to meet and exchange ideas'; and biennial meetings that 'bring together all members of the Human Variome Project Community . . . to meet and "cross-pollinate" '. As the article notes, 'These well-attended and popular meetings allow the Human Variome Project to coordinate solutions to issues that cross Working Group borders' (http://www.human-variomeproject.org/index.php/meetings) (accessed 9 April 2010). An HVP news bulletin published in February 2010 announces, among other items, that the third upcoming HVP meeting (hosted in May 2010) was granted UNESCO patronage and that the meeting was to be held at the headquarters of UNESCO in Paris. It also noted that Australia was creating a data repository called the Australian Human Variome Database that would mark the commencement of the Australian Node of the Project and that this would 'act as a pilot system/model for the establishment of other country nodes'. Further, the project team had commenced on the regulatory structure and ethical statement that 'will assist in the ongoing sustainability and oversight of the project' (Human Variome Project Bulletin, 2010: 3).

With its repository of information on publications, news items, pilot projects, and weblinks to groups of related interests, the HVP website reveals that this is an ambitious undertaking comprising a well developed network of scientists with powerful patronage who are developing an innovation that is contributing to shaping of our understandings of health, illness, and identity. However, the visibility of this project among the population at large appears to be low, with little apparent opportunity for lay publics to have a say on the direction, value, and likely implications of the project. The website reveals nothing about the project partners' views on the ethical and social implications of innovations and how information is likely to be applied in practice, especially given the global de-regulation of biomedicine and the contemporary focus on 'care of the self'. It seems to be simply assumed that such a resource will be useful, with no evident reflection on the question of who will ultimately benefit from the information, whether it may be used in ways unanticipated, and how it will likely change conceptions of health and illness and society. The fact that the Human Genetics Commission—a purportedly unbiased advisory group of the UK government—perceives this as a potentially useful resource for patients, without offering critical comment on its utility and implications, is worrying and underlies its limited conception of the 'ethical' issues raised by such databases.

Lab Tests Online: Taking the Laboratory to the Community

The development of the website Lab Tests Online, comprising a number of national versions that pay cognisance to differences in culture and language, typifies the global trend towards the creation of mechanisms for enhancing consumer choice in healthcare. These Lab Tests Online websites are portrayed as a kind of democratisation of science involving the taking of the laboratory to the community (as one web-linked document describes it, the 'disseminated laboratory') (http://www.acb.org.uk/docs/ACBCarterSubmission.pdf) (accessed 16 April 2010). Close inspection of details on the sponsors of the websites reveals that these are strongly supported by governments and biotechnology and pharmaceutical companies. The websites typically include an introductory blurb trumpeting the benefits for the individual consumer (e.g. 'Lab Tests Online-UK has been designed to help you, the patient, to understand the many clinical laboratory tests that are used in diagnosis and treatment') (http://www.labtestsonline.org.uk/) (accessed 16 April 2010), prominent webpage links to 'Tests', 'Conditions/Diseases', and 'Screening', as well as links to 'Understanding your tests', 'Inside the lab', 'In the news', 'Article index', 'About this site', and a section for feedback ('Send us your comments'). It is explained that one can use the site for diagnosing and managing conditions and diseases and search for tests directly. That is, the websites are presented as a 'one-stop shop' for consumers who are presumed to be confronted with a mass of potentially confusing information on conditions, diseases, and treatments in the global healthcare marketplace. The portrayal of genetic tests as but one kind of test among the many available serves to 'normalise' tests as a healthcare tool. There is no discussion of the potential problems and uncertainties posed by the use of particular tests. The tests arguably gain legitimacy through their endorsement by governments and health organisations as well as private industry. For example, NHS Direct, NHS24 (a health information and advice service for Scotland) and Health-EU ('the official public health portal of the European Union') appear on the homepage of the UK version of Lab Tests Online (see http://www.labtestsonline.org.uk/) (accessed 16 April 2010).

As the UK version of the website explains,

> Lab Tests Online-UK is the product of a unique collaboration among professional societies representing the clinical laboratory community. The concept was originally conceived by the American Association for Clinical Chemistry (AACC), and launched in the United States of America as Lab Tests Online. The potential value of the site was identified by the Association for Clinical Biochemistry (ACB), and through generous funding by a grant from the Health Foundation, it has been able to licence the material from Lab Tests Online and create a site tailored to the UK public. (http://www.labtestsonline.org.uk/site/acb_partners.html) (accessed 16 April, 2010)

In its submission to the 2006 Carter review on the modernisation and improvement of NHS pathology services in the UK, *Modernising Pathology:*

Building a Service Responsive to Patients, the Association for Clinical Biochemistry (ACB) outlines its commitment to the provision of community-based laboratory services, its support for the concept of the 'Disseminated Laboratory' where laboratory medicine 'should straddle primary/second/tertiary care services' and its 'Overall vision' which includes a strong emphasis on 'access and patient choice' through the Lab Tests Online website and other media (ACB, 2006: 3). The altruistic intent behind the websites is frequently emphasised. For example, the homepage of the US version states:

> Lab Tests Online has been designed to help you, as a patient or family caregiver, to better understand the many clinical lab tests that are part of routine care as well as diagnosis and treatment of a broad range of conditions and diseases. (http://www.labtestsonline.org/site/index.html) (accessed 16 April 2010)

The language of empowerment infuses these websites, with numerous references to how the information provided will enable patients to take control of their health. This is clearly evident, for example, on the 'Home test' link of the US Lab Tests Online website. Under the heading 'With home testing, consumers take charge of their health', it is noted:

> *You can reap the benefits of home testing—convenience, privacy, control—as long as you educate yourself about the potential tradeoffs.*
>   If you've been to the drugstore lately, you may have noticed an increase in the number of medical tests you can use in the privacy of your own home. Advances in testing technology—and changing attitudes towards patients' responsibility for their own health care—have made home testing a worldwide, billion-dollar-and-growing market. In fact, the word 'patient' itself is gradually disappearing—people like you, who used to think of themselves as patients, are now hearing themselves called 'consumers' who are taking charge of their own health care. (Emphases in original; http://www.labtestsonline.org/understanding/features/hometesting.html) (accessed 23 April 2010)

It is interesting to note how the marketers of these tests, noting the shift in emphasis in healthcare from 'patients' to 'consumers', have deployed the language of consumerism and individualism to advance their marketing goals (i.e. the reference to 'convenience, privacy, control'). Although this particular website is marketed to audiences in the US which has a particularly individualistic approach to healthcare, similar language can be found on other national websites, reflecting the globalisation of the rhetoric of consumerism that characterises contemporary healthcare in general.

The language of collaboration and partnerships also heavily infuses these websites and linked documents. For example, the above, ACB submission talks extensively about 'partnerships' within laboratory medicine in the UK—within the NHS, with the diagnostics industry, with community pharmacies, and with the independent sector—and how this will be utilised to realise 'new service delivery approaches' that will both improve quality and 'value for money' (ACB, 2006: 9). Whereas those who developed this particular website and the other national versions are no doubt motivated by the best of intentions, such appeals to altruism obscure the substantial interests that stand to benefit professionally and/or commercially either directly or indirectly from the diffusion of testing culture. These include interests in blood banking, pathology organizations and services, and the clinical laboratory industry (see, e.g. Lab Tests Europa Online) (http://www.labtestsonline.info/eu.html) (accessed 16 April 2010). A number of these interests, it should be noted, see it as their role to lobby governments to affect changes in healthcare in line with their particular vision of future healthcare. As the visibility of such websites increases and more and more information on testing becomes available, it will become increasingly difficult to regulate this field because a 'path dependency' will be created, whereby early decisions create momentum for and 'lock in' later decisions along a narrow track (see, e.g. Djelic and Quack, 2007).

EUGENICS BY THE BACK DOOR?

A question which arises with the proliferation of genetic tests directly to consumers via the internet and the clinic is whether there is a potential for unintended 'eugenic' outcomes to arise from their routine use (see Chapter 1). Some writers see the 'new' genetics—which encompasses both innovations in preventive medicine and public health (Petersen and Bunton, 2002)—as inherently 'eugenic'. For example, Jeremy Rifkin writes, 'The new genetic engineering tools are, by definition, eugenics instruments', and whereas the old eugenics was shaped by political ideology, fear, and hate, 'the new eugenics is spurred by market forces and consumer desire' (1998: 128). Social scientists such as Suzanne Anker and Dorothy Nelkin have commented that, although some of the eugenic programs of the past have been discredited, eugenic thinking persists (2004: 132). Over the last two decades or more and corresponding with increased research in genetics and interest in its applications in healthcare, such concern has intensified. Troy Duster and a number of other writers have warned of eugenics making its appearance through the 'backdoor' of screens, treatments, and therapies (Duster, 1990: x). A range of groups have expressed concern that the 'mainstreaming' and 'normalisation' of genetic tests and genetic-based treatments in healthcare may serve eugenic outcomes regardless of the intentions of those who use them. This includes feminists, disability groups, and ethnic minority groups who have suffered discrimination, including through the use of biomedical interventions to 'breed out' 'undesirable' groups (e.g. Nazi eugenics). By seeking to 'do the right thing' in health and exercising 'choice' through undertaking genetic tests, patients may unknowingly implicate themselves in eugenic programs. However, the question arises: is genetic testing inherently 'eugenic'?

To begin, we need to be clear about what is meant by 'eugenics'. As some writers have pointed out, there is no single universally agreed definition of 'eugenics'. Whether an outcome is judged to be 'eugenic' or not will depend on the definition adopted. Like many widely used terms, 'eugenics' is attributed multiple meanings and is often used rhetorically by those who object to particular ways of thinking or acting without reflection on the social contexts of its deployment. The concept of informed consent as applied in the context of the clinic and in research (e.g. biobanks), it is frequently argued, evolved in response to the eugenic programs of the Nazi period (see Chapter 1). In bioethics and social science discussions about new genetic innovations, there is rarely acknowledgement of the complex histories and diverse manifestations of 'eugenics'. The question of whether or not these are coercive is also open to contestation and dependent on whether one focuses on *aims/intentions* or on *outcomes/effects*. Further, there are contending meanings of 'coercion', which vary between different political traditions. Whereas classical liberals and contemporary (libertarian) conservatives view coercion as the deliberate interference of other human beings, for socialists and some strands of liberalism, a *situation* may also be coercive, such as being confronted with the medical and other costs of caring for a child who is severely disabled (Paul, 1998: 101). As Diana Paul observes, the word 'eugenic' has nasty connotations and serves to mobilise anxieties; however, the meaning of 'eugenic' is obscure and often reveals more about the user's attitudes than about the practices, policies, and intentions or consequences so labelled (1998: 109). Although often narrated as an official story of right-wing, reactionary thinkers and politicians, eugenics encompasses a complex range of ideas. Both Left and Right sides of politics have supported versions of 'eugenics', and eugenic programs have involved a diverse array of actors united only by the belief in the value of applying scientific precepts to society (Quine, 1996). I consider the question of whether the 'new genetics' is 'eugenic' and elaborate on the character and complexity of the various eugenic discourses elsewhere (see Petersen, 2007c).

Of the many social science conferences and workshops focusing on genetics that I have attended over the years, few sessions and presentations have focused on the histories or multiple meanings of eugenics. This is not to deny the excellent scholarship examining the cultural histories of eugenics movements and practices; however, my observation is that these rarely inform social science discussions about the character and implications of the 'new genetics'. If one is to be able to assess the potential implications of the widespread use of genetic tests in healthcare and lifestyle decision-making, it is important to be clear about what is meant by 'eugenics' and a 'eugenic' outcomes. Proponents and opponents of the new genetics use 'eugenics' for different rhetorical purposes: the former to distinguish 'it' from the positives of the new genetics, the latter to point to the dangers of the 'slip-pery slope'. However, a distinction is rarely made between the intentions, means, and outcomes of policies and practices, or to specify the underlying views of the human subject or the particular forms of citizenship associated with the 'new genetics' and 'eugenics', respectively. Particular conceptions of 'eugenics' and its history dominate the thinking of scholars with little evident awareness of how this shapes ethical protocols, such as informed consent and other practices that surround the use of genetic technologies. The identity of bioethics is strongly founded on a version of eugenics con-ceived as coercive practice, with the role for bioethicists being seen to be about developing means of 'protecting' or 'promoting' the autonomy of the individual largely through adherence to certain principles. However, eugen-ics so conceived sits uncomfortably with current healthcare philosophy and practice which is oriented to patient choice and empowerment.

Following the disability scholar and activist, Tom Shakespeare (1998), a useful distinction may be made between 'strong' eugenics and 'weak' eugenics, the former being a coercive form of practice which is imposed by authorities of the state and institutionalised in many countries at var-ious times through the twentieth century (i.e. not just Nazi Germany) involving the control of reproduction to create population-level improve-ments, and the latter entailing the promotion of 'technologies of repro-ductive selection via non-coercive individual choices' (1998: 669). As the line between genetic testing for health advancement and genetic testing to facilitate decisions about lifestyle and enhancement becomes less and less clear, parents may begin to routinely select embryos not just for reasons of 'health' but also for displaying genetic markers of socially desirable char-acteristics, such as tallness, intelligence, complexion, and so on (Petersen, 2007b: 91). The idea of 'germline' genetic engineering, which involves 'cleansing' the human gene pool of 'faulty' genes is supported by some scientists who argue that this will eliminate suffering and reduce health-care costs in the future (see, e.g. Stock and Campbell, 2000). During periods of economic hardship, as during the period of World Depression of the 1930s, many commentators from both the Left and Right argued for control over the human gene pool, for eliminating those groups that are considered to be economically unproductive and hence a social bur-den (Allen, 1996). Similar arguments, but in different guises, have since reappeared during periods of economic hardship. Whereas one of the arguments for genetic counselling is to guard against such genetic selec-tion by ensuring 'informed consent', as we shall see, its approach is not necessarily at odds with this outcome. Experts' concerns about the DTCA of genetic tests, described above, arise largely out of concern about the absence of the mediating role of the genetic counsellor and the exercise of the so-called 'non-directive' approach. However, the evidence indicates that 'non-directiveness' is far more difficult to realise in practice than the rhetoric suggests and is far from neutral in its effects.

'Non-directive' Genetic Counselling

'Non-directiveness' is seen as a defining feature of contemporary genetic counselling, a point underlined in textbooks written for students and practitioners in the profession as well as in statements articulated by professional associations. For example,

> Genetic counselling does not aim to prevent couples from having children with genetic diseases. Preventing genetic disorders, although important, is secondary to good clinical practice, which identifies couples at risk and by empathic counselling allows them to make their own informed choice about prenatal diagnosis, termination of pregnancy or other aspects of management. Informed choice without external coercion should distinguish medical genetics from eugenics, which has the contrary aim of improving the communal gene pool. (Harris, 1998: 335)

> As health care professionals, genetic counselors interpret and provide clear and comprehensive information about the risk of any medical condition that may have a genetic contribution. They ascertain the usefulness of genetic technologies for individuals and families and facilitate an informed decisionmaking process that elicits and respects the spectrum of personal beliefs and values that exist in society.(American Board of Genetic Counseling, Inc., 2008)

Terms such as 'empathetic counselling', 'informed choice', 'informed decisionmaking', and 'without external coercion' used in these statements and elsewhere, and references to how the practices of genetic counselling differ from other fields of medical intervention or 'eugenics' (e.g. its 'client-centred' emphasis and the promotion of 'informed consent'), are common in the professional literature. Underlying the use of these terms and pronouncements are particular conceptions of eugenics and of human subjectivity and freedom. In its self-representations, genetic counselling provides risk information to its 'clients' but withholds direct advice, enabling them to reach 'informed', voluntary decisions. This suggests that genetic knowledge can be value-neutral and that those who undergo counselling are independent, rational decision-makers. This denies the ways in which the framing of information (e.g. as a percentage (10%) or as a ratio (1 in 10)) and the contexts in which information is presented may shape its reception. It also disregards the influence of one's family, prescribed gender role, educational background, and cultural and religious communities on individual views and actions. Decisions are never unconstrained, and subjectivity is shaped by numerous factors. Individuals' views on genetic risk and genetic testing have been shown to be influenced by their prescribed gender roles and ethnic and religious communities and their felt sense of responsibility for others (see, e.g. Hallowell, 2009; Rehmann-Sutter, 2009). A grow-ing number of genetic counsellors have questioned the desirability of the non-directive approach and have wondered whether it can be realised in practice. Nevertheless, non-directiveness continues to be espoused as the guiding ideal of practice and rationale for the profession. Drawing on the dominant discourse of bioethics, which focusses on patient autonomy, genetic counsellors aim to create the autonomous client (Morrison, 2008). They do this by adopting an 'empathetic' approach, paying consideration to their 'clients'' perspectives, values, and beliefs, and yet ignoring their social relationships, the source of the social values that 'clients' express, and the constraints of context that shape their own practice (2008: 194–197). The idea that it is possible to be 'non-directive' underpins expressed concerns about DTCA, noted above. That is, it is assumed that if patients were pre-sented with an 'appropriate' level or kind of information by the counsellor, they will make a more 'rational' (presumed 'correct') decision. The ques-tion of what level or kind of information should be presented, by whom, and in what form and what might constitute a 'rational' or 'correct' deci-sion is rarely raised.

A point that is often overlooked in discussions about the provision of 'non-directive' genetic counselling is that the profession has not always adopted this approach as its ideal and indeed has been overtly directive in the past. Commentators often exhibit blindness to the history of genetic counselling and its practices, overlooking the fact that the first 'genetic counselling' centre, established in the 1920s and based at the Eugenics Record Office in Cold Spring, US, offered highly directive advice as to whether to marry. Consistent with the state-endorsed eugenic policies of the period, counsellors offered premarital, preconception, and post-con-ception hereditary advice with a view to the future of the gene pool, as well as the health of future offspring of their clients (Kenen, 1986: 174; see also Walker, 1998: 2–3). This extended into much of the post-World War II period. As recently as 1972, genetic counselling was described as a kind of 'preventive medicine' (see Leonard, et al., 1972: 437). Resta (1997) cites evidence showing that during the post-World War II period, parents were encouraged not to reproduce in cases where there was the potential for pro-ducing children with 'undesirable traits'. As Resta explains,

> Although they [geneticists] criticised eugenic *programs* that were based on racism and coercion, geneticists still felt that eugenic *goals* were compatible with the goals of genetic counselling. Nondirectiveness was a reaction to the methodology of eugenics, not its principles. (1997: p. 256; emphases in original)

Walker writes that in the late 1940s and 1950s, when few diagnostic tests were available, counselling was limited to offering families information, sympathy, and the option of avoiding child bearing. Geneticists assumed that 'rational' families for the most part would want to prevent occurrences of births with genetic disorders or disabilities (Walker, 1998: 3). This was congruent with the view of Francis Galton, the purported founder of eugenics, who believed that eugenics should be a kind of 'secular religion'; that is, citizens should act responsibly in their reproductive decision-making (Paul, 1998: 105). Increasingly from the 1970s, however, the provision of information alone has been seen as insufficient to 'empower' patients. The view is that patients must be assisted to become competent decision-makers and that the counsellor should assist them in reaching this goal through better understanding their social, cultural, educational, economic, emotional, and experiential circumstances and assisting them to better understand how various courses of action or events could affect them and their families (see, e.g. Walker, 1998: 8). The use of 'client' rather than 'patient' reflects this shift in emphasis: the former reflects subjects' capacity for independent decision-making and their readiness to put information to use, whereas the latter suggests passivity and limited capacity for independent decision-making (Brock, 1995: 158–159).

A growing body of critical scholarship has drawn attention to the constraints on 'non-directiveness' in the practices of genetic counselling posed by the ideological and political context and the power relations between counsellors and their 'clients' (e.g. Bosk, 1992; Chadwick, 1993 Clarke, 1991; Rapp, 1988). This has helped undermine the counselling profession's idealised version of its practice and raises questions about the implications of counsellors' continuing promotion of the ideal of non-directiveness. Among this growing literature is research showing how counsellors' views on genetics in reproductive decision-making ('repro-genetics') is shaped by culturally-specific interpretations of 'eugenics'. In a recent publication, reporting research based on interviews with Israeli and German genetic counsellors, Hashiloni-Dolev and Raz (2010), found significant differences in 'normative dispositions' towards Nazi period eugenics which has relevance for current practices of genetic counselling. While, as the authors emphasise, there is no single 'national' German or Israeli view on the significance of the Holocaust, the German counsellors were found to be more sensitive to the disability critique of repro-genetics than their Israeli counterparts (2010: 97). They interpreted Nazi history as having lessons for current bioethical dilemmas and as alerting them to the dangers of the revival of eugenic policies. Consequently, they tended to prioritise society's interests, in terms of maintaining genetic diversity, before individual interests. Israeli genetic counsellors, on the other hand, tended to view the Holocaust as a singular, anti-Semitic event and as irrelevant to their work, stressing the importance of genetic technologies for national strength and survival. As the authors conclude, these differences had direct implications for views on social responsibility towards disability:

Israeli respondents highlighted, in a characteristically straightforward manner, the conflict resulting from certain parents' insistence on carrying 'problematic' pregnancies to term, at the expense of burdening society. German respondents, in contrast, emphasized a different conflict—that between the individual wish to avoid having abnormal children and society's need for genetic diversity and tolerance. (Hashiloni-Dolev and Raz, 2010: 97)

While it is not known how and to what extent these different 'national' views shape practices in genetic counselling because the study is based on interviews alone, this research emphasises the danger in assuming that there is universal consensus about the social role of genetic counselling; namely, whether it should operate to protect the rights of 'the individual' or 'society'. It highlights the different interpretations of 'eugenic' and how this may shape counselling practice. In the case of Germany, views supporting genetic diversity can be assumed to offer some constraint on consumer-driven repro-genetics. In Israel, on the other hand, a more particularistic view on the Holocaust combined with a liberal emphasis on individual choice and market forces, the authors suggest, is more likely to promote a more commercially and consumer-driven repro-genetics (Hashiloni-Dolev and Raz, 2010: 98–99). Notwithstanding such national or cultural diversity in views and, potentially, practice, the growing availability of genetic tests on the internet and via the clinic would seem to be pointing towards an intensification of consumer-driven healthcare at the global level that has uncertain, long term socio-political implications. Further, like other areas of DTCA, this is a field that is difficult to regulate at the national level. In this context, the deployment of the current repertoire of bioethics concepts, particularly 'autonomy' and 'informed choice', can be criticised not only for diverting attention from the broader social justice implications of these developments but, more fundamentally, for helping to confirm their validity. In particular, the role of bioethics in adding legitimacy to the commodification of the body and life itself needs to be acknowledged. The implications of bioethics discourse in newly emerging areas of technology, especially nanotechnology, carries a similar danger, as I explain in the next chapter.

6 Governing Uncertainty
The Politics of Nanoethics

The vision of the National Nanotechnology Initiative (NNI) is a future in which the ability to understand and control matter at the nanoscale leads to a revolution in technology and industry that benefits society.

(National Science and Technology Council, 2007: 9)

Nanotechnology is more than an exciting new technology. It represents a whole new method of manufacturing, which achieves control at the atomic scale. It is better described as a collection of technologies which are genuinely "disruptive"—that is, they will render many existing technologies and processes obsolete and create entirely new types of products. . . . Nanotechnology has been described as a new industrial revolution.

(House of Commons Select Committee on Science and Technology, 2004: 5)

As with the biotechnologies considered in earlier chapters, nanotechnologies are the subject of great expectations. However, with nanotechnologies optimism is especially high, a commonly held view being (reflected above) that they will contribute to or constitute a revolution. Comprising phenomena at the molecular (billionth of a metre) level, these technologies are seen to embody unique properties that will create novel opportunities in a range of areas. Current and predicted future technology applications span the fields of biomedicine, engineering, information and communication systems, energy storage, and the production of a growing array of consumer goods, including paints and personal care products such as sunscreens and cosmetics (Woodrow Wilson International Centre for Scholars, 2010).

Despite these optimistic predictions, as the experience of other earlier technologies reveals, there is nothing inevitable about the path of development of nanotechnologies which will be subject to the influence of various economic, political, and social factors. Technologies often develop in ways unimagined and have unintended impacts—both positive and negative. Further, as commentators frequently observe, the applications of nanotechnologies will be contingent on their convergence with other technologies, particularly genetic technologies, digital technologies, and neuro-technologies whose development paths are also unpredictable. Uncertainty surrounds the nature, extent, and timing of this convergence. Experience of other technologies also shows that there are always 'winners' and 'losers' from innovations. however, the question of who will likely 'win' or 'lose' from nanotechnologies cannot be foreseen ahead of applications.

The theme of uncertainty has dominated discussions about nanotechnologies since the field began to gain public prominence in the early years of the twenty-first century and has proved challenging for those who seek to promote its benefits as well as those who are concerned with its governance. The question of what constitutes nanotechnologies, whether the technologies are genuinely novel or existing technologies that had been 'rebranded', how and when they will be applied and with what effects have been hotly debated among biophysical scientists, policymakers, ethicists, and social scientists. Uncertainty surrounds the attendant biophysical risks and, indeed, whether the existing means for measuring risks are adequate. The science itself is contested, with scientists divided on the question of how to 'characterise' nanotechnologies ('the size, shape, distribution, mechanical and chemical properties') and how technologies may be fabricated (e.g. via 'bottom up' or 'top down' techniques, or a convergence of both) (see, e.g. The Royal Society and The Royal Academy of Engineering, 2004: viii, 25–29). The many uncertainties that surrounded this field and the lack of an established language and set of metaphors for describing its boundaries and content have created the conditions for the flourishing of a rich fictional imagery and debate concerning the character and implications of nanotechnologies. Research on media coverage of nanotechnologies during the early phase of their public prominence in the UK reveals the existence of extensive fictional imagery in news stories, with depictions drawn from popular works such as Michael Chrichton's *Prey* (Anderson, et al., 2005, 2009). The pervasiveness of this strong fictional imagery within the public discourse on nanotechnologies has enabled those who seek to promote their preferred visions of the future much scope to create their own narratives about the significance of innovations.

This chapter examines the role played by bioethics in the governance of nanotechnologies. I argue that ethics and specifically bioethics, along with philosophy and the social sciences, have been called upon to help engender legitimacy and establish consent for this field of research and development (R&D) that has already achieved considerable momentum propelled by high expectations. Ethics plays this role through contributing a particular language, namely, the language of democratic participation and citizen rights, and a specific repertoire of techniques for rationally managing 'public opinion'. With new and emergent biotechnologies, ethics is ascribed a key role in managing attendant biophysical and personal harms or 'risks'. Risk, by definition, is calculable (being based upon probabilistic information), and hence viewed as subject to control through mechanisms of preventive intervention. But how does 'ethics' operate in the field of nanotechnologies which entails so much uncertainty (i.e. where the risks are incalculable)? I contend that ethics discourse plays a key role in translating the uncertainties of nanotechnologies into the familiar language of risk, thus making them *governable*.

THE EXPECTATION EFFECT

As with other technologies considered in earlier chapters, for example, genetic technologies and stem cell technologies, nanotechnologies have been seen as crucial to the burgeoning information-based economy, with many governments in recent years jumping on the 'nanotech bandwagon' in order to gain a competitive edge in the field. However, as noted, the expectations surrounding nanotechnologies are especially high given the range of purported applications with direct economic 'spinoffs'. The level of government and scientific commitment is substantial. In the US, the UK, Europe, Asia, and Australia, the field of nanoscience and nanotechnologies has attracted considerable research funding in recent years (e.g. The Royal Society and The Royal Academy of Engineering, 2004: 1), the expectation being that the findings of science *will* 'translate' into usable technologies that will advance economic development in the future. Government and industry reports often cite the expected economic benefits in justifying investment in R&D in the field. For example, in the 2008–2009 Annual Report of the Australian Office of Nanotechnology, it is stated that 'Worldwide revenues from products incorporating nanotechnology are projected to reach US$2.84 trillion by 2015, driven by an expected exponential growth in commercialisation successes in the healthcare and electronics sectors' (Australian Office of Nanotechnology, 2009: 8). Further, it notes that, 'By the end of 2008 nearly $40 billion had been invested by governments in nanotechnology research', and that 'There is little evidence that the Global Financial Crisis [of 2008–9] has affected nanotechnology spending by Governments' (Australian Office of Nanotechnology, 2009: 9).

Nanotechnology R&D has come to constitute a huge industry in its own right, comprising numerous activities and actors located within diverse sectors, agencies, institutions, fields of practice, and/or disciplines, including biomedicine, materials science, information and communication technologies, policymaking, regulatory bodies/agencies, universities, research funding bodies, conference organisers, bioethics, philosophy, social science, legal scholarship, advertising, and public relations. In other words, many groups directly or indirectly contribute to and benefit from the establishment and maintenance of the high expectations that pertain to nanotechnologies. Not being limited to a particular sector like healthcare, the financial, institutional, and personal career investments in this field are more extensive than are those in many other fields of technology considered in the previous chapters.

In the European Union, the network of agencies, professional organisations, and individual experts that support nanotechnology-related initiatives is extensive. A scan of the website of Nanoforum.org (grandiosely subtitled 'European nanotechnology gateway') and its *Nanoforum* e-newsletter (published since 2005), which provides a source of news and events relevant to nanotechnologies, provides insight into the range of activities and actor groups within this field. According to its website blurb, Nanoforum was originally funded by the European Commission under its Framework 5 Program but now (since July 2007) operates as a European Economic Interest Grouping. It goes on to note, Nanoforum 'will continue to provide news items from across the EU including information from projects and organizations, and can now offer customers a powerful advertising and dissemination service through its network of over 15,000 registered users and a website that attracts over 100,000 visits, 400,000 page impressions and 900,000 hits each month' (http://www.nanoforum.org/nf06~struktur ~0~modul~loadin~folder~140~.html?) (accessed 2 July 2010). In July 2010, the website included links to an extensive list of news items (3,984), organisations (2,500), events (1,802), calls and programs (169), and publications (494). The nanotech conference sector alone is big business, with a Nanofair and a number of nanomedicine conferences advertised in 2010. A map of organisations throughout Europe highlights the vast number of industries throughout Europe producing goods involving nanoparticles or nanomaterials. A 'Nano Education Tree' provides a 'one-stop shop' of information for those wishing to learn more about the purported applications of nanotechnologies, a history of its development ('Nano Timeline'), and its 'Societal impacts'. This 'Tree' includes various links which inform viewers of the substantial benefits that nanotechnologies will deliver. One, 'Modern life', states, 'Virtually every area of modern life will be touched and improved upon by nanotechnology in the coming decades' (http://www.nanoforum. org/educationtree/modernlife/modernlife.htm; bold in original) (accessed 2 July 2010).

The Nanoforum newsletters provide further insight into the range of activities and actors involved in the field of nanotechnologies. For example, the 54th *NanoForum* newsletter, published in June 2010, includes among its items an announcement of the publication of the ObservatoryNANO newsletter; details on the NanoPharm Mission 2010 (a three-day conference providing 'a proven format for providing new business relationships'); an announcement of the launching of '12 new projects in the Dutch societal dialogue on nanotechnology, in addition to the 21 projects which have been ongoing since January 2010'; an article on Dutch MPs' involvement in 'a fact finding mission on nanomedicine'; and an article announcing that the European Parliament would be updating the regulation on 'novel foods' which since 1997 has excluded nanotechnologies (http://www.nanoforum.org/nf0 6~modul~loadin~folder~8074~sent~~step~~.html?) (accessed 2 July 2010).

In the US, the National Nanotechnology Initiative (NNI), established in 2001 'to coordinate Federal nanotechnology research and development' (http://www.nano.gov/html/about/home_about.html)   (accessed   2   July

2010), involves a vast number of departments and agencies participating in nanotechnology research in that country. The list is extensive, including: the Department of Defense, the Department of Education, National Institutes of Health, Bureau of Industry and Security, National Institute of Food and Agriculture, Environmental Protection Agency, Department of Justice, Department of Labor, Department of State, Nuclear Regulatory Commission, and National Aeronautics and Space Administration (NASA). The number of Centres and 'Networks of Excellence' supporting multidisciplinary teams from different sectors, including industry, academic, and government laboratories and spanning domains such as health, defense, materials science, engineering, nanobiotechnology, and environmental protection, is vast (http://www.nano.gov/html/centers/nnicenters.html) (accessed 2 July 2010).

With climate change and economic and environmental sustainability firmly on the political agenda across the world and concerns about the limits of miniaturisation enabled by silicon chip technologies in computing, many governments look to nanotechnologies to help resolve urgent problems of economic and environmental sustainability and energy security. Nanotechnology-based industries based on diverse sectors have burgeoned in recent years, supported by substantial government funding initiatives, such as the US' NNI, the European Union's Framework Programs, and the UK's research council funding schemes (e.g. Engineering and Physical Sciences Research Council). Underlying these expectations for nanotechnologies is a conception of economic growth that is mostly unquestioned but increasingly is assessed as *unsustainable* (see, e.g. Jackson, 2009). That is, the current model of consumption-led growth is recognised as having contributed to the depletion of non-renewable resources, high levels of pollution, growing inequality, and an overall decline in community wellbeing. However, emergent problems are seen as mostly amenable to the technological 'quick fix', particularly through making economies and supportive information and social systems more productive, efficient, and 'flexible'. The expectation of 'miniaturisation' in electronics and medicine and the manufacture of 'harder', 'lighter' materials in engineering, enabled by nanotechnologies, can be understood in this light.

The nanotechnology field is one where very high expectations about technologies (in this case, in relation to resolving pressing economic, social, and environmental problems), combined with great uncertainties about the direction, timing, and implications of their development, presents acute challenges for governance, particularly in regards to managing uncertain public responses to the technologies. In order to establish a stable environment for investment in new technologies, and hence growth, authorities need to establish and maintain public support over the longer term by gaining public consent and legitimacy for research, especially through developing appropriate regulatory mechanisms. In diverse forums, authorities have expressed concerns about the dangers of 'adverse' public responses to the field of nanoscience and nanotechnologies, often making reference to earlier technologies, such as GM crops and food. In the UK and some other countries, scientists and policymakers often speak of the dangers of a 'GM-style backlash' against what is seen as a highly promising new field. The field of nanotechnologies is consequently seen by many as a 'testing ground' for developing a new approach to the science–society relationship.

Governing Public Response

While scientists and policymakers have taken stock of earlier technology controversies, public responses to nanotechnologies are especially difficult to predict because, unlike biomedical technologies, the 'user' groups are various, and the particular uses of technologies cannot be easily specified at the outset given the uncertainties about the direction and outcomes of technology convergence. Thus far, these technologies have not been the focus for citizen organisation and activism as have genetics and stem cell technologies. Patient groups have not formed around conditions defined by an emergent nanomedicine, and, if public surveys are a guide, the level of public visibility and knowledge of the field is low in many countries (European Commission, 2010: 25; Market Attitude Research Services, 2008; The Royal Society and The Royal Academy of Engineering, 2004: 59–60). Notwithstanding the existence of some local organisations and networks developed on the basis of the promises and expectations of nanotechnologies, the political dynamics of this field are mostly different from that of genetics, personalised medicines, and stem cell research. Scientists, science policymakers, and others who support and/or have a stake in the field cannot take public consent and trust for granted. There is currently no 'nanotechnology citizenship' equivalent to 'biological citizenship' (Rose and Novas, 2005), characterised by active consumers who have developed identities around particular conditions and who seek to gain access to promising technologies to treat them (see Chapter 5). Further, despite the stated opposition of some NGOs (e.g. Friends of the Earth in Australia, the ETC Group in Canada (which has called for a global moratorium on nano-based products) (see, e.g. Bird, 2010; ETC Group, 2003) and concerns of unions and consumer groups about certain aspects of nanotechnologies, particularly the safety of nanoparticles (Australian Council of Trade Unions, 2009; Ishizu, et al., 2008: 251–252), the sites of potential resistance to technologies are unclear. Whereas some national surveys suggest that citizens are generally optimistic about nanotechnologies in general (e.g. MARS, 2008; The Royal Society and The Royal Academy of Engineering, 2004), until *particular* applications of technologies are widely established and associated concerns addressed citizen support cannot be taken as given. Further, the potential long-term 'downsides' of particular technologies cannot be foreseen. As the experience of other emerging technologies (e.g. GM crops and food, cloning) have underlined, uncertainty is a precondition for the generation of fear. This is an area where governance is of an anticipatory or pre-emptive kind, with an uncertain, potentially fearful, hostile public response being constructed as a 'risk' along with bio-physical risks, to be managed by the regulatory tool of 'public engagement' or 'dialogue'. Later in the chapter I examine how 'engagement' has operated in practice, highlighting its socio-political implications. I go on to argue for the importance of acknowledging uncertainty in public deliberation—indeed making it central to the adoption of a more democratic approach to nanotechnology innovation. To begin, however, I believe it is useful to provide some comments on 'nanoethics'—a field whose parameters are yet to be settled but which takes a strong cue and a number of ideas from the more established bioethics.

THE RISE OF 'NANOETHICS'

Among scholars concerned with the ethical implications of nanotechnologies, considerable debate has focused on the question of whether these technologies present novel implications requiring new approaches or whether the issues that emerge are of a kind and order similar to other areas of technological development, say genetics or stem cell technologies that can be addressed through established normative frameworks. In line with this concern, a new journal, *Nanoethics*, was launched in 2007, purportedly to cast light on the above and other related questions posed by nanotechnologies. In his Editorial introduction to the first volume of this journal, John Weckert asks,

> Is there a branch of applied ethics, nanoethics, similar to, for example, bioethics or computer ethics? It can be argued, quite plausibly, that nanotechnology does not raise any new ethical issues; it is just more of the same. Anything that might be considered an issue in nanoethics is an issue in some other branch of applied ethics, so what is the point of nanoethics? (2007: 1)

Contributors to that volume and subsequent volumes have explored whether there can be and should be a dedicated nanoethics and, if so, how one might 'do' nanoethics, especially given that there appear to be few current applications raising concerns, apart from the nanotoxicity of nanoparticles, and that all that can be done is based upon prediction alone. Whereas discussions reveal diverse views on the contributions of and prospects for nanoethics, writers broadly agree that a key challenge in assessing the implications of nanotechnologies arises from nanotechnologies' *emergent* and thus *unpredictable* character. Because nanoethics is speculative rather than concerned with issues arising from established applications, commentators have been heavily reliant on science fictional imagery which, as Wood, et al. (2007) found, tends to polarise around radical visions (utopian or dystopian) of how technologies will develop (see also Wood, et al., 2008).

Writing in the first issue of *Nanoethics*, Sparrow (2007) has drawn attention to the contradictory narratives that characterise the nanotechnology
field. On the one hand, nanotechnologies are presented as revolutionary, in
that they represent a radical break from earlier technologies and hence pose
problems for their regulation. On the other hand, they are portrayed as
familiar, in that they are continuous with previous technologies—perhaps
existing for decades—and hence can be regulated through established regulatory tools. Sparrow suggests that these depictions work rhetorically, to
either support the position that the field presents novel, useful applications
or, alternatively, in instances where there is fear about the implications of
technologies, to render the technologies familiar and thus harmless. If they
are depicted as incremental extensions of existing technologies, then, in
Sparrow's view, this makes it plausible to claim that their development is
inevitable (2007: 62). Sparrow notes that the narratives used to describe
nanotechnologies are themselves familiar in that they pertain to issues that
are common to all new and emergent technologies, such as environmental
risks, social impacts, and political questions (2007: 66). He warns that
because the issues are likely to be different for different technologies, one
should be wary of thinking about the future in terms of these narratives,
but instead develop new, more complex ones that are 'sensitive to the speci-
ficities of the particular technologies' (2007: 66).

In another article in the same issue of this journal, Swierstra and Ripp
argue that 'there is little specific to nanotechnology that would warrant
the prefix "nano" ', which is 'different from the case of bioethics, where
aspects and issues derived from living creatures are a shared starting point'
(2007: 3). As they argue, it is an 'umbrella term' which encompasses a host
of 'heterogenous technologies' with different uses which, by virtue of being
'enabling'—that is, 'making existing technologies smaller and faster'—may
make ethical issues (e.g. privacy and new ICT) 'more pressing but not necessarily different in kind' (2007: 3). In the authors' view, there are certain
patterns of moral argumentation pertaining to nanotechnologies which are
general to new and emerging science and technology (NEST) in general.
While there may be some nano-specific issues, such as those posed by the
size of the technologies, new and emerging technologies in general disrupt
established 'moral routines' because they involve novelty and unknowns
('about what the technology might become and do') (2007: 6). This novelty
is stressed by the technologies' promoters who wish to attract financial,
political, and moral support (2007: 6).

Another author, van de Poel, writing in a later volume, acknowledges
the inadequacy of the current framing of ethical debates within the nanotechnology field, which he argues are both 'seriously misconceived' (2008:
28). As he argues, in discussions about the role of ethics in nanotechnology,
ethics has been conceived in two main ways: either as a means to establish
limits on the development of technologies and nanotechnology research
and development, or as a means of facilitating their development by easing

public fears and lack of acceptance (2008: 28). According to van de Poel, the problem with both these positions is that not enough is known about what nanotechnology will bring to society and what the ethical issues will be. These positions are based upon speculation rather than knowledge of how technologies are actually evolving. Further, he argues, they both focus on communication—to convey the dangers of technologies, or to overcome 'knowledge deficits', respectively—which represents a misunderstanding of the contributions of ethics, which as a philosophical discipline should be about analysing issues and providing arguments about the 'pros' and 'cons' of certain positions. Finally, he contends, the issues are complex, and the role of ethics cannot be reduced to providing simple yes and no responses. In light of the limitations of current ethical responses in this field, van de Poel concludes, ethical issues should be addressed during the early stages of nanotechnology development (2008: 34–36). Issues should be analysed in a 'real world context' *as they emerge* rather than viewed in the abstract. He proposes a 'network approach' utilised in science and technology studies which analyses the contributions of various actors and their 'different problem definitions'. The insights developed through attention to 'the empirical facts of the situation', he argues, can immediately inform research and development decisions (van de Poel, 2008: 36).

The difficulty faced by ethicists, philosophers, and social scientists in articulating a critical response to technologies that are emergent and involve considerable uncertainty is underlined when one examines contributions to *nanobioethics*, a sub-field within nanoethics that has been delineated to reflect on the particular challenges posed by the biomedical, biotechnological, and agrifood applications of nanotechnology (see Malsch and Hvidtfelt-Nielsen, 2010; Pavlopoulos, et al., 2010). This field, which is dominated by European philosophical perspectives, is portrayed as offering (according to one account) 'an ongoing ethical assessment of the norms governing the practices of new technologies' (2010: 69). The novelty of the field, however, has been contested, with some evaluating it as 'a continuation of earlier biomedical ethics debates' (Malsch and Hvidtfelt-Nielsen, 2010: 23). The field has been nurtured via substantial European Commission funding that has been devoted to ethical reflection on nanotechnologies as part of the effort to pave the way for expected innovations, rather than being part of a concerted endeavour to involve diverse publics in critical reflection and in efforts to shape the priorities of the field. Published documents in the field are notably *uncritical*, in that they provide little reflection on issues of politics, power, and knowledge and on the economic interests and expectations that sustain nanoscience and nanotechnologies. Further, there has been little analysis of the factors that may facilitate or impede citizen involvement in key decisions affecting science and technology policies. The ascribed role for nanobioethics is that of offering distanced observation on developments *as they evolve* rather than facilitating citizens' active intervention so as to shape science and technology, which would imply a radically different conception of citizenship and the science–society relationship than that which currently exists and openness to questioning the assumptions of science.

Nanoethics as a 'Toolkit'

ObservatoryNano, funded by the EU's Framework 7 program for four years from 1 April 2008 in order to 'to support European decision-makers with information and analysis on developments in nanoscience and nanotechnology', for example, offers a 'toolkit' for reflecting on the social and ethical impacts of nanotechnologies (http://www.observatorynano.eu/project/) (accessed 5 July 2010). According to its Foreword, this 'helps, and perhaps guides, ethical thinking by offering concepts, notions, and methods for an application to practical cases' (Pavlopoulos, et al., 2010: 3). The 'toolkit' includes a series of boxed questions pertaining to various nanotechnology applications, for example, in the areas of nanobiotechnology, nanomedicine, food and cosmetics, information and communication technology, and the military, followed by explanatory boxes elaborating on pertinent key concepts (e.g. 'tanshumanism', 'health and disease', 'public and private'). Under a subheading, 'What can ethics do for nanotechnology?', the authors describe the likely transformations wrought by nanotechnologies:

> Sometimes we say that technology has made our lives better, but sometimes we aren't so enthusiastic. Technological changes can cause an abrupt change in social history but can also go unnoticed. It is difficult to predict with certainty what a given technology will do. So what should our attitude be with regard to this uncertainty? What shall we do? How to make a judgment? This is where ethics enters. Scientists no longer live in an Ivory Tower. Society wants its share of influence on what comes out of the laboratory, because it knows that the uses of technology will have consequences for its own well-being. Whether the nanotechnologist wants it or not, he is not alone: his fellow citizens are interested in his work and want to control future applications. The responsibility of the scientist de facto exceeds his own sphere of action. How to deal with this situation? How to make sense of the scientist's new role? This is where ethics enters. . . . (Pavlopoulos, et al., 2010: 7)

Despite scholarly contributions such as this, which appear to be signalling a new, critical field of deliberation, few writers have challenged the dominant, determinist conception of technologies; namely, that technologies are evolving or will evolve in a certain direction according to their own inherent logic. Rather, as presented here, nanoethics offers little more than a 'checklist' of questions or issues to be considered when assessing technologies. 'Technologies' are portrayed as 'pre-social' and as being an independent 'driver' of economic and social change, and as thus are assumed to be unaffected by the operations of politics and power. In the above passage, for instance, 'ethics" contributions are framed in terms of assisting in (abstract) deliberations on how 'society' may obtain 'its share of influence on what comes out of the laboratory' rather than analysing the processes that enable laboratory work to be accomplished and 'translated' into 'technologies' (Latour and Woolgar, 1986). As Wood, et al., note, in their analysis of the literature on the social and economic implications of nanotechnologies published between July 2003 and April 2006 (a follow up to an earlier 2003 study), many depictions of nanotechnologies reflect an overly optimistic perspective of how technologies will develop. As they note, 'There is an almost "effortless" view of science and technology underlying many of the accounts as if new technologies flow easily from the daily life of the bench scientist to the market place' (2007: 4).

This optimistic depiction of science and technology in analyses of the ethical and social implications of nanotechnologies mirrors that found in many policy and science reports on nanotechnologies published in recent years. For example, The Royal Society and The Royal Academy of Engineering report, *Nanoscience and Nanotechnologies: Opportunities and Uncertainties* (2004), which has come to be widely regarded as being a 'landmark' report in the field, provides a range of scenarios for future technology development. The following description, taken from this report (under a heading 'Visions for the future'), is an example:

> There are strong drivers to reduce tolerances in engineering, including miniaturisation, improved wear and reliability characteristics, automated assembly and greater interchangeability, reduced waste and requirement for re-work. As the trend towards miniaturisation continues, research and the industrial application of energy beam processing methods will increase, driven in particular by the electronics and computer industries. (2004: 30)

In this depiction, technological innovation is inherent within science itself—'drivers to reduce tolerances in engineering . . . ', etc., which serves to further 'drive' innovation through investment in research and development. 'Miniaturisation' is seen to obey a technological imperative. There is no reference to the economic, political, and social factors underpinning technological change. These include the economic pressure to reduce costs of production through the adoption of new labour-saving technologies and to generate profits through the creation of an ever-growing number of consumer items such as computers and other electronic devices that are designed to become quickly obsolete. Further, the potential social and environmental costs of this anticipated path of technological innovation are given scant attention.

THE DISCOURSE OF BENEFITS AND RISKS

Debate about the ethical and social significance of nanotechnologies has been constrained thus far by a prevailing discourse of 'benefits and risks', with discussion about the latter focussed particularly on the dangers posed by manufactured nanoparticles and nanotubes (i.e. the biophysical harms). This is evident, for example, in The Royal Society and The Royal Academy of Engineering report (2004: Chapter 5) and in media coverage of nanotechnologies in Australia and the UK (e.g. Harrison, 2009; Sample, 2004). As Wood, et al, observe, 'analysis of nanotechnology's potential effect on society tends to focus narrowly on the possible toxicity of nanoparticles to human and environmental health, detracting attention from discussions of nanotechnology's wider impacts' (2007: 17). This narrow framing of the implications of the field is not surprising when one considers that much, if not most, research on the social and ethical implications of nanotechnologies is funded as an adjunct to large science programs and projects, in the same way that ethical, legal, and social implications (ELSI) research is funded under the Human Genome Project, rather than existing as a legitimate field of research in its own right. For example, the major funders of nano-scale science and technologies, the US National Science Foundation and the European Parliament's Sixth and Seventh Framework Programs assigned budgets for social and ethical research, which has helped fuel the growth of nanoethics and nano ELSA/ELSI (Kjølberg and Wickson, 2007: 89). As I explain later in the chapter, however, *public reaction* is also implicitly constructed as a 'risk', with governance efforts being oriented to managing this risk.

Kjølberg and Wickson's (2007) work provides insight into the preoccupations of nano ELSA/ELSI thus far. As they found in their review of research on the social and ethical implications of nanotechnologies (published in English language only), it is only since 2000 that one could refer to a body of work focussing on the social and ethical implications of nanotechnologies. Further, 2004 and 2006 were 'standout years', perhaps reflecting, Kjølberg and Wickson suggest, a number of events in previous years. These include the publication of Michael Crichton's *Prey* in 2002, which is widely credited with raising public interest in the social and ethical implications of nanotechnologies; the passing of the Nanotechnology Research and Development Act by the US Senate; widespread publicity in the UK surrounding Prince Charles' purported comments on the social implications of nanotechnologies; the publication of a report by the The Royal Society and The Royal Academy of Engineering which explored the economic, social, and ethical challenges of nanotechnologies; the ETC Group's report demanding a moratorium on nanotechnology development; and the European Parliament's first seminar on the societal impacts of nanotechnologies (2007: 92). The close alignment of nanoethics research with research that is funded by governments that have a stated commitment to nanotechnology-based innovations very likely explains the strong focus on governance issues (decision-making, regulation, legislation, public engagement) found by Kjølberg and Wickson. As the authors note, 60 percent of the articles focused directly or indirectly on governance issues, which was the largest theme found within the literature on the social and economic implications of nanotechnologies (Kjølberg and Wickson, 2007: 93). Although from these data it is difficult to assess the proportion of such research that is funded by government research programs, as is evident from other fields of research, investigation of social and ethical questions tends to *follow* rather than inform funding priorities. Research that is critical of the direction of government policy is far less likely to be supported than research which is congruent with policy priorities. Research councils in fact often stipulate government research priorities in their grant programs which serves to steer programs of research and limits investigation of a more critical, 'blue-skies' nature.

The UK's Experience

The narrow framing of debate on nanotechnologies via a discourse of 'benefits and risks', with the latter including biophysical harms and social responses, can be seen in the Royal Society and The Royal Academy of Engineering report which, as mentioned, is widely considered to be a 'landmark' document in this field in recent years. While this framing is evident throughout the report, it is pronounced in certain chapters, particularly those focussing on 'social and ethical issues' (Chapter 6) and 'Stakeholder and public dialogue' (Chapter 7), where 'risks' associated with particular applications are highlighted, either explicitly or implicitly in discussion. In the former, relatively brief (seven-page) chapter, it is noted that whereas some social and ethical concerns 'may not be new or unique' to this field, 'some nanotechnologies will raise significant social and ethical concerns' and that these concerns rarely become matters of concern merely as a result of the underlying science or engineering but rather as a consequence of their specific applications (The Royal Society and The Royal Academy of Engineering, 2004: 51). The report goes on to acknowledge that predicting these applications, and hence likely concerns, is difficult with nanotechnologies; indeed more so than with biotechnologies (for the reasons outlined above). Social and ethical issues include potential threats to civil liberties (e.g. covert surveillance through use of biosensors and chips), development of human enhancement applications, and military applications. As the report observes, there are likely to be 'winners' and 'losers' from the introduction of nanotechnologies (as there are with other technologies), such as the displacement of employment between different sectors of the economy and an intensification of the gap between rich countries and poor countries in terms of their abilities to develop and exploit the opportunities created by technologies, leading to a so-called 'nano-divide' (2004: 52).

These potential scenarios are underpinned by an uncritical view of technology development, with purported deleterious outcomes being implicitly framed as 'risks' in need of diligent 'watchfulness'. For example, in the concluding paragraphs, it is noted, 'There is a need to monitor future applications of nanotechnologies to determine whether they will raise social and ethical impacts that have not been anticipated in this report'. Further, there is the recommendation that 'the consideration of ethical and social implications of advanced technologies (such as nanotechnologies) should form part of the formal training of all research students and staff working in these areas' (2004: 57). In this discourse, 'ethics' is ascribed a vigilance role and positioned as an adjunct to the science, rather than a critical, agenda-setting role, in helping to forge deliberation and action on more substantive questions such as whether particular technologies should be developed at all, whether the kind of society presumed and shaped by technologies is desired, what interests lie behind the science and how this may shape the technologies, and so on.

The following chapter (Chapter 7) of The Royal Society and The Royal Academy of Engineering report focuses on one of the key issues that has been identified in relation to the governance of nanotechnologies in the UK, Europe, Australia, and a number of other countries, namely, how to identify and support potentially beneficial applications of technologies while managing the attendant risks. It is in this chapter that the construction of *public response as a risk* is most clearly evident. As it has been articulated in this report (and in other science and policy reports), the aim is to establish a 'responsible development' of nanotechnologies (2004: xii, 83–84, 87)—a task that is to be achieved via 'public engagement and dialogue' (2004: 87). As the report's working party noted, on the basis of previous research and its own survey and two in-depth qualitative workshops, public awareness of nanotechnologies is low (only 29 percent of the respondents were aware of the term 'nanotechnology' and only 19 percent could offer any form of definition) (2004: 59). Given the uncertainties surrounding this field and potential for this to shape public perception of the issues, as happened with earlier technologies such as nuclear energy and GM crops and food, a key aim is to 'educate' and 'inform' 'the public' during the early phase of the emergence of the technology. That is, it is important to view nanotechnologies as an "upstream' issue (2004: 64). This approach, it is argued, will not only improve trust in institutions, but help resolve 'potential conflicts in advance' and 'improve decision quality' (2004: 66).

Context of the 'Upstream' Engagement Approach

It is clear from the above that, in the UK 'public engagement' had from the outset been conceived as a means to engender consent and legitimacy for nanotechnology innovations—to 'build trust' that is seen to have declined in the wake of a number of previous technology controversies, particularly concerning GM—rather than as a means of potentially changing the priorities or direction of science and technology. The absence of such engagement—the tendency to only consult 'downstream' about the risks—it was argued, had contributed to adverse public reactions to technologies. In the lead up to The Royal Society and The Royal Academy of Engineering report, UK policymakers expressed concerns that the UK had lost the lead it had established in the nanotechnology field in the 1980s and had been overtaken by other countries, particularly the US, Germany, and Switzerland (see House of Commons Select Committee on Science and Technology, 2004: Chapter 2, paragraph 16). Other countries in Europe and Asia and the US were observed to be investing heavily in the field. For example, as noted, the US announced its NNI in 2001, with funding of $495 million, doubling its then federal research in nanotechnology (National Science and Technology Council, 2000: 13). The Royal Society and The Royal Academy of Engineering report, which was commissioned by the UK government, confirmed that nanoscience and nanotechnologies had already achieved considerable momentum and that technologies were expected to 'deliver' many of their benefits within coming years. It was seen as important, therefore, that the UK act promptly to capitalise on this promise. In this context, social scientists, philosophers, and bioethicists have been recruited to help develop mechanisms for 'engagement'. Scholars who have moved into the field of nanotechnologies from other fields, particularly bioethics, history and philosophy of science, sociology of science, and science and technology studies (STS), have brought their 'skill sets' and assumptions from these areas, and many have been willing and active participants in this new, engagement agenda by either researching and writing about engagement/dialogue matters in their professional journals or being actively involved in developing engagement-in-practice 'experiments' that have been generously funded in a number of countries. In recent years, and particularly since the publication of The Royal Society and The Royal Academy of Engineering report in 2004, 'public engagement' has become a large industry its own right, especially within Europe, with many scholars building careers within the field.

THE POLITICS OF 'PUBLIC ENGAGEMENT'

'Early' or 'upstream' public engagement has found a broad receptive audience, its language resonating among both policymakers seeking an urgent solution to the aforementioned governance conundrum and communities of scholars and NGOs concerned with the implications of new technologies. Whereas it seems to have originally taken root in the UK and Europe, the language of engagement has been picked up to varying degrees elsewhere, such as in Australia under its National Enabling Technologies Strategy and related initiatives (see http://www.innovation.gov.au/Industry/Nano-technology/Pages/public_forums.aspx) (accessed 7 July 2010). Drawing extensively on the discourse of participation found in fields such as the new public health and community development (e.g. Petersen, 1994; Petersen and Lupton, 1996), 'public engagement' appears to be self-evidently about the empowerment of citizens in relation to science and technology. One of the arguments for 'upstream' public engagement is that this will allow for debate about issues before views become fixed—the presumption being that many citizens will have the opportunity to participate in shaping policies at an early stage of technology development. For example, in The Royal Society and The Royal Academy of Engineering report, it is argued that one of the reasons that nanotechnologies should be viewed as an 'upstream' issue is that:

> many of the significant decisions that will affect the future trajectory of the technology, concerning research funding and R&D infrastructure, have yet to be made. . . . One driver of the current concerns among NGOs . . . is a scepticism over whether the technology will be shaped in such a way that its outcomes will genuinely benefit society, the environment and people (particularly in the developing world) as is sometimes claimed. A timely and very broad-based debate might therefore focus upon which trajectories are more or less desirable, and who should be the ultimate beneficiaries of public sector investment in R&D, before deeply entrenched or polarised positions appear. . . . (2004: 64)

This report goes on to cite evidence showing that a failure to consult the public on biotechnology 'led to several difficulties in the regulatory process' and that 'the problematic issues that heralded the advent of biotechnology in the 1970s and 1980s did not go away' (The Royal Society and The Royal Academy of Engineering, 2004: 64). The above comments, expressing concerns about 'entrenched or polarised positions', suggest that the adoption of this approach is underpinned by fears about public resistance to technologies rather than by a commitment to developing mechanisms for involving citizens in shaping the direction of technological development. Taking their cue from earlier technology controversies, scientists, policymakers, and supporters of nanotechnologies have learnt a lesson on the dangers posed by a fearful, 'mistrustful', and resistive 'public' and have sought means to engage citizens early in the cycle of technology innovation to engender consent and legitimacy for developments.

As noted in Chapters 1 and 2, 'public engagement' *in practice* thus far has tended to reinvent the so-called public deficit model of public understanding involving a conventional relationship between experts (who are presumed to possess valued knowledge) and 'the public' (which is assumed to be 'ignorant' or 'unaware' of 'the facts'). Like other widely used terms such as 'community', 'public engagement' is rarely defined and obscures more than it reveals about the workings of power in contemporary societies. In particular, it denies the power of experts to establish and control the 'engagement' agenda, including the conditions under which this occurs, the timing of its occurrence, and the makeup of the groups who are to be 'engaged'. The reference to a unitary 'public' denies the multiplicity of views and positions on the necessity for engagement and the most appropriate means and the desirable outcomes of its realisation. There is no acknowledgement of the difficulties of developing truly inclusive mechanisms of participation and of the imperatives that are implied for those who are involved. Few discussions about 'public engagement' or 'public dialogue' specify in any detail how the outcomes of the envisaged engagement will feed into policies and programs and how and when expected changes should be evaluated. According to some writers, 'engagement' should be centrally about enhancing 'informed' decisions among consumers of technologies (see, e.g. European Commission, 2010; Lenk and Biller-Andorno, 2006; von Schomberg and Davies, 2010). This reflects adherence to the bioethics principle of respect for autonomy, the assumption being that information derived through 'engagement' will 'empower' consumers to make 'rational choices' about technologies. It also suggests, as Nisbet and Scheufele argue, that 'science literacy is both the problem and the solution to societal conflicts' (2009: 1). Among bioethics, philosophers, and social scientists involved in public engagement on nanotechnologies, there has been little discussion about what 'informed' decision-making about technologies might mean when the field involves so many uncertainties.

'ENGAGEMENT' IN PRACTICE

'Engagement' has become embedded in many nanotechnology research programs across the world, although it has taken somewhat different forms and has had varying emphases in different jurisdictions. Insofar as it is meaningful to generalise at the global level, the UK and Europe have been characterised as taking a more 'precautionary' approach than the US and Asia in matters of science regulation. Whereas the 'precautionary principle' is used as a rationale for precautionary action, in practice different jurisdictions provide their own interpretations of the principle, with 'strong' and 'weak' versions evident in Europe (Malsch and Hvidtfelt-Nielsen, 2009: 7–11). The US NNI, established in 2001 to coordinate federal nanotechnology R&D, includes 'public engagement' as one of the specified fields of interest within its ELSI program and has established a Nanotechnology Public Engagement and Communications Working Group (http://www. nano.gov/html/society/societal_dimensions.html) (accessed 21 June 2010). However, what this means in practice is unclear. The information that is readily available in the public domain focuses on 'educating the public about nanotechnology', which is undertaken by organisations such as The Nanoscale Informal Science Education Network, comprising researchers and science educators (including science museums), which produces a website of information 'for teachers, students or anyone interested in nanoscience and the many potential nanotechnology applications' (http://www.nano.gov/html/edu/home_edu.html#NISE) (accessed 21 June 2010).

The UK's Economic and Social Research Council (ESRC) has funded a few projects that either focus on public engagement or the public representations of nanotechnologies (including one in which the author was involved), and the Government has funded some projects under its 'Sciencewise' program and Copus Grant Scheme, including Small Talk, Nanodialogues, and the Nanotechnology Engagement Group (Gavelin, et al., 2007: 7) A number of other 'public engagement' projects, not funded by government, including the NanoJury UK, Nanologue, and Nanoforum, were also conducted in the years following the publication of The Royal Society and The Royal Academy of Engineering report (i.e. 2004–2006). In Europe, since 2004, the European Commission has published a number of reports and funded a series of projects on nanotechnologies which have emphasised the importance of communication and dialogue (see European Commission, 2010). A recently published report, *Understanding Public Debate on Nanotechnologies: Options for Framing Public Policy*, edited by von Schomberg and Davies (2010), highlights some of the range of projects recently funded by the European Commission under its 'Science in Society' program. The cross-European project, 'Deepening ethical engagement and participation in emerging technologies' (DEEPEN), involving Germany, the UK, the Netherlands, and Portugal, funded by the EU, is but one among a number of such projects. A scan of the recently published European Commission report, *Communicating Nanotechnology: Why, To Whom, Saying What and How?*, highlights strong commitment to the 'engagement' approach. This presents a 'toolkit' of engagement/dialogue strategies and practices, outlining a vast range of materials, technologies, and tactics that have been or can be used to 'engage' different audiences, including school children, policymakers, NGOs, business, industry, funding bodies, and insurers (European Commission, 2010).

In Asia, little effort has been devoted to 'public engagement', with that which has occurred being mainly about 'educating the public' in order to engender understanding and acceptance of technologies. In Japan, where scientists played a major role in the early stages of development of nanotechnologies (1960s), policymakers were reluctant until 2004 to encourage debate about the societal implications of nanotechnologies for fear that this would provide an obstacle to the development of technologies (Ishizu, et al., 2008: 234). In that year, it launched a forum titled 'Nanotechnology and Society', which aimed to encourage 'debate about the impact of nanotechnology on society, and to create broad network of researchers and government to consider the societal impacts of nanotechnology in Japan' (2008: 236). However, this forum, which entailed a series of stakeholder meetings and a symposium, appears to have been limited to researchers, policymakers, government departments, business interests, journalists, and educational institutions (2008: 236). The forum was followed by the launch of 'Research project on facilitation of public acceptance of nanotechnology', in 2005, which again was expert-dominated, with participants organised into working groups and workshops to discuss 'the societal implications of nanotechnology and ways to address them' (2008: 237). The general approach to engagement in Japan seems to be about engendering *public acceptance* of nanotechnologies, correcting 'misunderstandings' about 'risk', and encouraging citizens to actively inform themselves about the technologies (2008: 252–253). As 'public engagement is currently framed, there appears to be little scope for citizens to shape the agenda of debate or direction of science and technological development, which is seen in that country (which has suffered a long period of recession) as crucial for economic development.

The Purported 'Crisis of Trust'

In the UK, interest in 'public engagement' in relation to nanotechnologies arose against a background of a purported 'crisis of public trust' following in the wake of a series of failed government responses to earlier technology innovations. As noted, The Royal Society and The Royal Academy of Engineering report strongly emphasised 'engagement', and around the same time of the launch of this report, the London-based think-tank, Demos, released its pamphlet, *See-Through Science*, which called for public engagement in relation to nanotechnology (Wilsdon and Willis, 2004). Interestingly, this was published in partnership with the environmental think-tank Green Alliance, the Environmental Agency, and the Royal Society, reflecting the broad consensus about the importance of such an approach among public and private organisations representing science and environmental groups. A 2007 report describing a series of UK-based public engagement projects, *Democratic Technologies?*, published by Involve, a not-for-profit organisation focusing on public engagement, provides a common account of the background to this crisis in the UK (Gavelin, et al., 2007). This includes the Bovine Spongiform Encephalitis (BSE) controversy, which highlighted widespread public concerns about the role of science in regulatory processes and the use of science-based GM technology in UK food production (2007: 2). This was followed by government enquiries and reports from science groups, such as The Royal Society, which emphasised the need for a new approach to risk and risk communication that involved the public in risk decision-making, including risk assessment and regulation. Henceforth, the argument goes, public values were to be considered along with science facts in deliberating upon science issues and setting standards as part of the regulatory process (Gavelin, et al., 2007: 3–4).

From the outset, 'public engagement' was seen to assist in overcoming a perceived problem of 'loss of public trust' and 'deep ambivalence' about science that was assumed to be reflected in hostile public responses to earlier technological controversies. By moving public engagement 'upstream', it was argued, there would be more scope for citizens to be involved in examining the purposes of scientific research and technological innovation—that is, in shaping technological trajectories—rather than debating technologies 'downstream' when public scepticism 'brought about through poor engagement and dialogue on issues of concern' may hold back technological development (Gavelin, et al., 2007: 4). The report refers approvingly to the Demos' pamphlet, mentioned above, which outlines the arguments for 'upstream' public engagement:

> The pamphlet argues that in the case of the GM controversy, public dialogue was entered into at a point when it was too late to influence the development of the technology. Thus, the public engagement activities overlooked a core element of the controversy: the fact that people were protesting not only against the products and technologies that were emerging on farms and supermarket shelves across the country, but also against the underlying conditions and assumptions that had allowed these products and technologies to be developed in the first place. (Gavelin, et al., 2007: 4–5)

As the report notes, the claim that the trajectory of technology development itself was at issue is based upon understandings of public attitudes to new technologies derived from social science research. Public concerns focused on questions such as how technologies are used, the social distribution of risks and benefits, and the capacities of governments to respond to unforeseen future consequences, rather than just scientifically defined risks to human health or the environment (Gavelin, et al., 2007: 5).

Despite an avowed concern with these 'social' questions, the UK's program of public engagement has focussed resolutely on the technologies themselves—both the issues that they raise and institutional responses to public concerns and aspirations that emerge—and on developing mechanisms for establishing and maintaining public confidence and trust in the technologies. (For an overview of the program's stated 'aspirations', see Gavelin, et al., 2007: 6–7.) Notwithstanding the rhetoric, relatively little effort has been expended in exploring what engagement/dialogue might mean in practice, including the desired means and ends of citizen participation in relation to science and technology. As it happened, the public engagement 'projects' and 'activities'—dubbed by some commentators (e.g. in the Gavelin, et al. (2007) report) as 'experiments'—have been mostly expert-driven, organised, and orchestrated by university researchers, representatives of government agencies, research councils, and business organisations (e.g. The Guardian, Unilever), with some involvement of NGOs (e.g. Greenpeace UK) and an assortment of other groups including science museums (e.g. Ecsite-UK) and science festivals (e.g. Cheltenham Science Festival). They were also initiatives of short duration, ranging from two months to two years in duration (no doubt conforming with budget cycles), comprising mostly a series of 'one-off' events (e.g. citizens juries, workshops). The value of such mechanisms in terms of changing the power relations between citizens and science and advancing a more democratic culture of science and technology has been extensively critiqued (see, e.g. Irwin and Michael, 2003). However, these critiques seem to have been overlooked or ignored by sponsors, who have tended to revert to what can be described as a predominant technocratic 'top-down' approach.

Impediments to 'Engagement'

Significantly, interviews undertaken with scientists and policymakers by the Nanotechnology Engagement Group identified a number of institutional and cultural impediments to undertaking 'public engagement', including a lack of incentives, resources, time, and support and the absence of experience, training, and outlook among professionals in relation to engagement strategies and practices (Gavelin, et al., 2007: 62–72). (This concurs with Katz, et al.'s (2009) assessment of Australia's initial 'public engagement' efforts, described later in this chapter.) Different groups were found to have different expectations of public engagement on nanotechnologies.

> We found an expectation on the part of the public participants that the findings and recommendations would be used by decisionmakers to inform nanotechnology policy. To date, there is little evidence that this happens. We have also found that decision-makers assumed that effective public engagement should be aligned with policy needs and provide outputs that fit neatly to policy-making structures. (Gavelin, et al., 2007: 62)

Revealingly, the recommendations arising from the Involve report are addressed almost solely to decision-makers in government and science: 'Government should spend money on nanotechnologies provided that priority is given to funding research and developments that contribute to a wider social good, such as new medical innovations and sustainable technologies'; 'Government should continue to identify the potential risks of nanotechnologies and nanomaterials, and create new regulation and laws for labelling based on such research'; 'Government should take steps to ensure that the governance and funding of nanotechnologies is made more transparent'; 'Scientific institutions to formally recognise public engagement'; and 'Science-funding bodies to stress the importance of dialogue-focused public engagement, alongside one-way engagement approaches such as public lectures' (Gavelin, et al., 2007: xi–xii). There is no challenge to the idea that 'public engagement' should be located within the sphere of expert deliberation and control. None of the recommendations addresses broad social priorities—that is, go beyond the sphere of science—which would call for a fundamental rethinking about the role of science and technology in society, including about how knowledge is produced, which fields of science should be supported, how any expected benefits can be equitably distributed, and so on. If citizens are to help shape the priorities of science, as the discourse of 'upstream public engagement' suggests, one needs to remain open to the prospect that citizens may reject certain kinds of research, prioritise areas not currently valued by scientists, or place a higher value on other kinds of activity. Further, there needs to be acknowledgement of the fact that science itself as it is currently practiced arguably *contributes* to social exclusion rather than social inclusion.

Australia's Experience

In Australia, the language of public engagement has more recently infiltrated government policy, with programs thus far centred on education and raising awareness of the field. For example, the National Nanotechnology Strategy announced in May 2007 by the then Liberal Government included a 'public awareness and engagement program', and the Australian Office of Nanotechnology, which ceased operations in June 2009, included a team of 'public awareness officers'. The stated aim of the 'public awareness and engagement program' was to 'raise awareness and develop knowledge of the opportunities and potential of nanotechnology and to encourage an informed debate based on balanced and factual information over two years' (Australian Office of Nanotechnology, 2009: 20). This program included the launch of an online educational resource, the development of a series of publications, and various 'community engagement activities', such as a 'Social Inclusion and Engagement Workshop', involving various stakeholder communities, that 'identified areas for engagement and principles to underpin engagement activities'; a 'nanodialogue' on occupational health and safety issues, and a second on nanotechnology and food 'to scope the issues that need addressing'; and a 'Webinar' on industrial uses of nanotechnology (2009: 21). The subsequent National Enabling Technologies Strategy, launched in February 2010 by the Labour government 'to provide a framework for the responsible development of enabling technologies such as nanotechnology and biotechnology—and other technologies as they emerge in Australia' (Australian Office of Nanotechnology, 2009: 6)—also includes 'public awareness and community engagement' in its remit. Its range of activities encompass the continued funding and expansion of an 'information and outreach service', TechNYou, based at University of Melbourne, a planned series of community engagement events, involving meetings with key stakeholders (NGOs, researchers, government bodies, and business) and 'information fact sheets' available via the web, and links to 'educational websites', research reports, and media releases (http://www. innovation.gov.au/Industry/Nanotechnology/Pages/PublicAwarenessan dEngagement.aspx) (accessed 10 July 2010).

Preceding these government initiatives, in 2004, Australia's Common-wealth Scientific and Industrial Research Organisation (CSIRO) funded and conducted two public engagement workshops focusing on nanotech-nologies (Katz, et al., 2005, 2009; Mee, et al., 2004). The first, which was located in a regional location (Bendigo), involved a range of stakeholders (twenty-two in total) 'in order to investigate how members of the public evaluate nanotechnology' and also to 'evaluate the range of social values, or criteria, that participants drew on in their assessments' (Katz, et al., 2009: 535). The second, held in Melbourne City, involved a seventeen-member Citizens' Panel comprising experts drawn from various contexts (commercial, environmental impact, regulatory, social impacts, ethics) and lay citizens who responded to advertisements posed in local newspapers and other forums and explored a range of issues related to nanotechnolo-gies, such as health and safety aspects, regulation, ownership, and control. These have been mainly expert-driven events of short duration (one-day) and with limited lay input. As with the majority of other 'public engage-ment' endeavours, discussed earlier, the primary focus of the events was on the technologies and their impacts and was limited by its stakeholder focus; that is, involving groups of experts with a declared interest in the area and self-selected members of 'the public'. Significantly, the social sci-entists who were involved in organising these events have acknowledged the limitations of both events in light of the uncertainties surrounding the definition of 'nanotechnology', the emergent character of the field, and the lack of clarity surrounding the parameters of the 'debate' or engagement process, especially in terms of the relative power of the different partici-pants (Katz, et al., 2009: 540). As they argue, there are 'barriers' involved in moving towards public engagement which derive from 'current research governance structures and levels of resourcing' (2009: 540). There is no positive incentive to undertake public engagement and no recognition of its value. Indeed, notwithstanding nanoscientists' 'in principle' commitments to engagement, there are concrete *disincentives* in terms of resourcing and people's competencies. A key problem in seeking to realise the ideals of public engagement are the impediments posed by organisational structures and cultures within which knowledge production occurs (2009: 541)

CONCLUSION

Despite its democratic overtones, the discourse of 'public engagement' that is promoted in many countries that are committed to developing nano-technologies has not signalled a significant shift in assumptions about the character and role of science and technology in society and constructions of citizenship. In practice, 'public engagement' has relied on traditional methodological approaches that are congruent with the conceptual model of the public understanding of science that has been substantially critiqued in recent years (Kurath and Gisler, 2009: 569). Science and technology are seen as separate from society, and thus unaffected by politics and power, and as essentially benevolent. In many cases, science communication involves little more than 'one-off' efforts of short duration to 'educate' or increase the 'awareness' of 'the public' about nanotechnology issues. Among ethicists, philosophers, and social scientists, there has been little reflection on the political, institutional, and cultural factors that impede the realisation of a more participatory approach to science and technology. The effect of 'public engagement' discourse and practice has been to de-politicise questions such as whether particular technologies should be supported and whether they assist in effecting *social changes* of a kind desired, who decides which applications are to be developed, and who is likely to 'win' and 'lose' from particular applications. Ethics and bioethics specifically, along with the social sciences, have been recruited to help resolve an acute problem of governance, namely, how to achieve support for nanotechnologies for which there are high expectations and considerable political support but many uncertainties. By contributing the inclusive language of citizen participation and citizen rights which provides a veneer of democratic inclusion and by focussing attention on a narrow range of issues and neglecting substantive questions, ethics and the social sciences have helped lend legitimacy to nanotechnology R&D. Further, scholars who write in the field of nanoethics have tended to have a restricted conception of their own potential contributions—as distanced observers on innovations as they evolve rather than as active participants in helping to shape the knowledge and direction of science and technology.

Whereas uncertainty has been framed as a problem for governance in the field of nanotechnologies, as with other new and emergent technologies, it can be seen alternatively as a focal point in initiating discussion about desired futures, about how and which technologies may assist in realising these futures, and about the likely consequences of pursuing certain technological innovations. There is a need to reflect on the role of expectations in shaping developments and on whether undue optimism has led to and may in the future lead to investment in technologies that fail to live up to expectations or that have unforeseen or dangerous consequences or that incur costs disproportionate to their benefits. In particular, the model of growth that underpins the high hopes for technological 'fixes' to complex politico-economic and socio-cultural problems needs critical scrutiny. Unfortunately, ethics and bioethics and the social sciences thus far have been unable to offer such an analysis, and one needs to look elsewhere in developing alternative perspectives on the field. It is with this in mind that the next and final chapter considers the future of bioethics and normative frameworks *beyond* bioethics.

7 Beyond Bioethics

> One egg, one embryo, one adult—normality. But a bokanovskified
> egg will bud, will proliferate, will divide. From eight to ninety-six
> buds, and every bud will grow into a perfectly formed embryo, and
> every embryo into a full-sized adult. Making ninety-six human beings
> grow where only one grew before. Progress.
>
> (Huxley, 1955: 17; orig. 1932)

The idea of being able to control and perfect nature has long been a theme
in science fiction and other popular imagery. In 1932, Aldous Huxley's
novel, *Brave New World*, portrayed a future where science is applied for
improving the quality of life and achieving universal happiness. In the story,
scientific knowledge is the 'highest good', serving 'Man' in overcoming the
constraints of nature, in order to alleviate human suffering, which had been
previously seen as 'God-given' or pre-ordained. Progress is measured by
the ability to manufacture life. Huxley's story has proved highly insight-
ful, with efforts to control 'nature' through processes of standardisation,
normalisation, and industrialisation, characterising how the life sciences
are practiced today.

While this and other widely read dystopian stories, most notably Mary
Shelley's *Frankenstein* (1985; orig. 1818), sound a warning about the dangers of 'tampering with nature', belief in the promises of science, and in
particular its potential to control processes of life to solve problems of
human suffering and thus achieve universal happiness, seems unshakeable.
The persistence of this belief was dramatically underlined in May 2010
with the announcement that researchers at the J Craig Venter Institute had
'successfully constructed the first self-replicating, synthetic bacterial cell'
(Fatimathas, 2010). According to one news report, the 'breakthrough' had
the potential to 'digest oil from leaks and spills, or bacteria that consume
cholesterol and other dangerous substances in our bodies' and 'attack other
microbes that cause so much death and illness' (Samuel, 2010). Citing the
lead scientist, Craig Venter, a second report notes 'This cell we've made is
not a miracle cell that's useful for anything, it is a proof of concept. But
the proof of concept was key, otherwise it is just speculation and science
fiction. This takes us across that border, into a new world' (Fatimathas,
2010). With a growing chorus of predictions such as this that science fiction is quickly translating into science fact—that the vision of absolute control over nature is on the verge of being fulfilled—there is an urgent need
to develop new critical perspectives on the claims that surround biotechnology developments. These should move beyond the utopian/dystopian

extremes that characterise many portrayals of biotechnologies, including those found in many bioethics writings, to offer a nuanced, *empirically-informed* assessment of developments and their contexts.

In the chapters, I have highlighted the significance of expectations in shaping science and technologies, which have tended to be uncritically accepted by bioethicists and others who have drawn upon bioethics concepts. Drawing on insights from sociology and science and technology studies, I have argued that these expectations are socially produced and supported through various reiterative practices, including the management of public representations. The vested interests involved in promoting particular visions of the future, it was noted in Chapter 2, are often substantial, including the pharmaceutical and biotechnology industries, science groups, governments, and patient groups, who, in sharing this vision, often work in collaboration, sometimes across national borders, frequently utilising the promotional techniques of public relations. I have emphasised the high optimism surrounding the contributions of biotechnologies to the 'bio-economy' which is being developed via a 'big science' model involving teams of researchers and many supportive institutions and personnel. This bio-economy presupposes and enables the formation of a new form of citizenship, 'biological citizenship', with biotechnologies seen to contribute to enhancing 'choice' among citizens who are presumed to play an active part in managing their own health. However, as I explained (Chapter 2), despite the potential of expectations to shape actions, there is nothing inevitable about the trajectory of the development of technologies which is subject to a variety of influences, particularly public resistance and changes in investment decisions resulting from, for example, financial crises. An understanding of these dynamics shaping expectations is crucial, I believe, in articulating normative approaches to biotechnologies that are appropriate to the times.

Whereas bioethics is frequently called upon to assist in identifying, debating, and responding to the value questions arising from new and emergent biotechnologies, the evidence considered in this book suggests that we need to look *beyond* bioethics. In Chapter 1, I noted the growing number of critiques of philosophy-based bioethics, arising in particular from sociology, feminist studies, and disability studies. I then went on to develop my own perspective on bioethics as it has been applied *in practice*. As I have argued, hopefully convincingly, bioethics has played a role in efforts to engender consent and legitimacy for new and emergent biotechnologies for which considerable political support and momentum has been established, often failing to offer critical insights into the development, character, and implications of technologies. Bioethics' concepts and principles and modes of analysis, I have suggested, have served a political function in helping to resolve the complex socio-political questions arising from existing, emergent, and imagined technologies that involve many uncertainties and conflicting positions and claims. By focusing on the technologies themselves and on only a limited array of issues and by offering a 'ready-made' repertoire of abstract concepts and principles to help address associated problems, bioethics has operated to restrict deliberation and action on issues. In short, it has served as a *tool of governance*. Having made my case, the question arises: what exactly lies beyond bioethics? For scholars and those concerned with the implications of the biosciences and biotechnologies, what should be the focus of critical attention? And, what might this mean practically for those who are concerned about the value questions arising from innovations?

The Democratisation of Science and Technology Development

As I have argued at various points in the book, one of the key limitations of bioethics in developing a critical response to the biosciences and biotechnologies is its failure to articulate a critical perspective on the science–society relationship. Science is seen as 'pre-social' or separate from 'society' and thus unaffected by social interests and power relations. 'Mainstream' bioethics offers no developed perspective on society and politics, and consequently bioethicists and those who draw on bioethics concepts are ill-equipped to offer critical insights into the dynamics of technological development and the socio-political implications arising from particular innovations. To use the language of science and technology studies, there is no recognition of the *co-construction* of science and society and of users and technologies (see, e.g. Oudshoorn and Pinch, 2003) (see Chapter 1). Bioethics has been primarily concerned with monitoring and reacting to technology issues that have emerged, are emerging, or are expected to emerge. Reflecting this emphasis, bioethics focuses on questions such as access to technologies, the protection of rights to privacy that are endangered by the use of technologies, the right to know versus the right not to know in the case of information derived from genetic testing, informed consent in testing or research involving technologies, and so on. This means that bioethicists' interventions are almost invariably reactive, or 'after the event', and consequently focussed on managing the individual and social impacts of innovations. Bioethics very often serves a kind of risk management function oriented to limiting deleterious consequences of past decisions that have 'locked in' innovations along a certain trajectory (Djelic and Quack, 2007). 'The public'—which tends to be conceived of as a unitary entity—is assumed to have no role in deliberating on and shaping the direction of science and technology.

In recent years, many science groups, science policymakers, and those researching and working in the fields of science communication, the sociology of science, and science and technology studies have questioned established conceptualisations of the relationship between science and society. Further, pressure groups such as Friends of the Earth, GeneWatch, and those working in independent think-tanks, such as UK's Demos, have called for communication approaches that advance the democratisation of science. The 'public understanding of science' approach, which assumes that the main task of science communication is to raise awareness and improve appreciation of science among an assumed unaware or 'ignorant' public, has been revealed as elitist and reinforcing the power relations that exist between scientists and non-scientists (see Chapter 2). Under this model, individuals are conceived as empty vessels that need to be 'filled up' by the infusion of more or better information. 'Science' tends to be portrayed as constituting established, relatively unproblematic 'facts'. In their efforts to overturn this orthodoxy, in some countries (e.g. the UK), the more enlightened science policymakers and science communicators have called for more early or 'upstream' public engagement on emerging technologies. This has been most clearly evident in relation to nanotechnologies, a field where, as explained in Chapter 6, a number of countries have undertaken 'public engagement' initiatives. Because they are seen as emergent, nanotechnologies are seen as providing a kind of test bed for 'upstream' engagement. However, as I argued, many or most of such initiatives thus far have arguably involved a reinvention of the so-called deficit model of public understanding which has dominated science communication for many decades. That is, it is assumed that public participation in science and technology is about raising public awareness or understanding of the science, rather than allowing scope for greater citizen participation in defining the overall direction of science and technology (Wynne, 2006). As Wynne comments, there has been little recognition of the *democratic deficit* that exists that excludes much of the population from debates about the aims and character of science and technology innovation. Scientists' own 'knowledge deficits', he argues, need to be exposed, debated, and subject to scrutiny. This includes interrogation of scientists' assumptions about 'society' and 'the public', about the social impacts of science, and about the means to democratising science. Other writers have argued that the complexities, ambiguities, and uncertainties of science should be an integral part of public discourse which is mostly lacking in debate (e.g. Cunningham-Burley, 2006).

While agreeing with these commentators, I would go further and argue that there needs to be a substantial rethinking of the aims and methods of 'public engagement' and conceptions of 'the public'—indeed, changing the very language in which this discussion about science and technology communication is couched. Surveys, focus groups, citizens' juries, and other forms of deliberation are arguably mechanisms for constructing and governing 'the public' rather than a means for democratising science and technology. Whereas there is currently much discussion among social scientists, bio-physical scientists, representatives of NGOs, and other commentators about the meanings of early or 'upstream' public engagement and the best mechanisms for its realisation, there has been relatively little reflection on exactly what is entailed in effecting a radically reconfigured science–society relationship. If one is serious about democratising science, one has to help nurture the conditions conducive to citizens playing an active role in shaping innovations at all stages of science and technology development and in deliberating on the value and likely implications of certain innovations. So-called 'constructive' technology assessment (CTA) (Schot and Rip, 1997) or 'real-time' technology assessment (Guston & Sarewitz, 2002), developed in science and technology studies, exemplifies approaches that seek to shift the focus away from assessing impacts of technologies once they have been developed to encouraging dialogue among and early interaction with various actors and making explicit the assumptions and values that underpin proposed innovations. A CTA approach seeks to 'broaden the design of new technologies' by allowing a 'feedback of TA activities into the actual construction of technology employing an appropriate array of strategies and tools' (Schot & Rip, 1997: 252). As Schot and Rip argue, despite the approach's promise, CTA is 'not a clearly defined area', and it may not become one, nor should it be a management tool (1997: 255). CTA will involve an element of societal learning, for example, by reflection on earlier technology innovations, and the development of appropriate mechanisms that allow feedback by actors (e.g. businesses, consumer groups, regulators) and methods for controlled experimentation in the early phase of technology development. As Schot and Rip point out, a core component of this approach is 'the attempt to anticipate effects or impacts of new technologies or new projects with a strong technological component' (1997: 257). Whereas in traditional technology assessment the technology or the project is taken as a given, for CTA emphasis is given to the dynamics of the assessment process, and the impacts are viewed as co-produced and established as technical change occurs (1997: 257). These impacts are viewed as actively sought and thus co-produced by various actors—by technology producers, governments, unions, and pressure groups with different degrees of access and power through the various stages of technology development (1997: 257). This is a radical proposition which suggests the need for new mechanisms for facilitating citizen engagement and/or the development of new forms of local decision-making where citizens decide for themselves what innovations should be supported and resourced—indeed, a participatory transnational politics of the kind advocated by Paul Ginsborg (2008: 102–111). An important question for deliberation concerns the ultimate ends of technologies—the kind of society that is envisaged—and the most appropriate means for achieving these ends.

Those concerned with the implications of the biosciences and biotechnologies could take an active role in influencing the processes that shape science and technology; in other words, facilitating global civil society. This suggests the need for more *anticipatory* normative approaches that involve the assessment of *potential realistic scenarios* of science and technology development. Whereas there will always be some uncertainties about what and how biotechnologies will develop, especially as they converge with other technologies (see Chapter 6), the positing of feasible scenarios, based upon a range of available empirical evidence (politico-economic, socio-cultural, technological), can assist in informing community deliberations about the likely and desirable direction of technological innovations. 'Horizon scanning' is already an accepted tool within governments and some non-government organisations. One example is the UK's Foresight programme (http://www.foresight.gov.uk/index.asp) (accessed 6 August 2010), which draws on a range of disciplines from the natural and social sciences to assist policymaking on science and technology. However, thus far, programs of this kind have tended to be closely wedded to specified governmental objectives, operating as a means to exploit the economic opportunities and manage the risks arising from likely future innovations, rather than serving as a mechanism to facilitate broad-based community deliberation on the 'rights and wrongs' of potential developments and assisting citizens to play an active role in shaping the future. What is needed, rather, is a kind of 'back casting', which entails mapping particular desirable scenarios for the future in light of known trends and 'drivers' of change and establishing the events and processes that would need to occur in order to realise that future (Dennis and Urry, 2009: 147–148). This approach draws attention away from technologies *per se* and towards a greater understanding of the economic, political, and social conditions that enable and constrain the development and shape the form of those technologies. Studies could explore the potential for citizens to shape the direction of the development of the biosciences and biotechnologies. Questions in need of exploration in this respect include: What scope is there for citizens to influence investment decisions pertaining to the biosciences and biotechnologies? How, if at all, does this differ in the case of private sector investment as opposed to public sector investment? To what extent and how can citizens shape the expectations that 'drive' innovations and supportive public and private investments and commitments? What mechanisms best facilitate citizen involvement in bioscience and biotechnology development? And, what factors impede involvement?

Attention to Issues of Global Justice

One of the major criticisms of 'mainstream' bioethics is its narrow focus on issues of immediate concern to the developed Western world (see Chapter 1). As noted, bioethics has been preoccupied with dilemmas arising from biomedical and biotechnology development as they have occurred in the rich countries of the West and North, especially the US, the UK, and Western and Northern Europe. These are contexts in which 'health' has been valorised to an extreme extent, where citizens are called upon to play an active role in self-governance through attention to the ideals of 'healthy living' (see Petersen, et al., 2010), where medicine is perceived to offer quick 'fixes' and where consumerism and individualism pervade healthcare and other aspects of life. Increasingly biomedicine and biotechnologies are oriented to redressing problems once seen as matters of chance or aesthetics (one's appearance, height, weight) or as an ordinary part of day-to-day living (e.g. feeling 'low' or 'unwell'). As many commentators have noted, in contemporary biomedicine, the line between 'treatment' and 'enhancement' has become blurred (e.g. Elliot 2003). Adherence to the ideal of perfectibility, envisaged in Huxley's Brave New World, underpins the growth of the market in enhancement products, including cosmetic surgery, and anti-aging medical treatments (Petersen and Seear, 2009). Bioethics offers a mirror of contemporary Western (particularly US) society and its preoccupations, values, biases, and idealised ways of living. Far less attention has been paid to pressing issues of concern in developing countries of the East and South, which are intimately connected to how people in the developed North and West live their lives. These include 'health' which tends to be framed more broadly and in collective terms rather than individualist terms: responses to the challenges of controlling HIV/AIDS (e.g. in Africa), water-borne and insect-borne diseases, poverty-related diseases, and the diseases and problems arising from globalisation itself and the take up of Western medicines, lifestyles, and consumption (e.g. high sugar, high fat intake, and alcohol and tobacco use). Differences in mortality and morbidity between countries of the developed and the developing world have long been documented. Even within developed countries, differences in life chances can vary considerably. For example, indigenous Australians are likely to live on average twenty years less than the rest of the population. Climate change and the economic crisis of 2008–2009 forced a number of issues of global justice on the policy agenda that are relevant to ethical debate pertaining to biomedicine and biotechnologies. These include food security, water shortages, energy security, changes in habitats, increases in disease, and growing poverty. In April 2009, the International Monetary Fund (IMF) announced that the economic crisis had resulted in 50 million people, particularly women and children, being driven into extreme poverty (http://www.imf.org/external/np/cm/2009/042609.htm) (accessed 13 July 2009). An Oxfam International report, *Suffering the Science: Climate Change, People, and Poverty*, published in 2009, underlined the severity of the problems confronting people in the developing world resulting from climate change. One of the most important is hunger due to the susceptibility of the world's staple crops such as a maize and rice to rising temperatures and 'to more unpredictably extreme seasons' (2009: 5). Other problems identified include water-borne and insect-borne diseases, increase in climate-related disasters, water shortages, climate-driven migration; and increase in conflicts between countries resulting from scarcity of water supplies (2009: 5–6). According to the authors, this necessitates that rich countries commit to reducing their contributions to global carbon emissions and assisting poor countries to cope with climate change (2009: 6). This demands shifts in economic and social systems and in patterns of consumption, energy use, and investment. The model of consumption-led growth, as measured by the Gross National Product (GNP), needs to be challenged given its contribution to the unsustainable exploitation of non-renewable resources and the generation of waste (e.g. Bock, 2010: Chapter 4; Jackson, 2009). A critical perspective on the biosciences and biotechnologies that placed issues of global justice centre stage would make stronger links between investment and spending priorities in the developed West—including on expensive drugs, tests, and treatments—and economic hardship and human-induced climate change in the developing world.recent reports have highlighted the close links among food scarcity, the growing costs of energy, economic crisis, and social and political unrest. The sharp increase in the price of wheat and maize in 2007 and 2008, to three times higher than the level at the beginning of 2003, and rice, to five times higher, were aspects of the inflation of food costs fuelled by a range of factors including domestic policies, the market structure, and transportation costs (von Braun, 2008: 10–11). Rising energy prices and the growing production of biofuels supported by policies of subsidization, which has led to a decrease in agricultural land, have been contributing factors to the food crisis (2008: 9). Notwithstanding a fall in the price of major cereals resulting from the economic slowdown towards the end of 2008, the costs of food remains high, and there are continuing doubts about the ability of the poor to achieve sustained access to adequate amounts of nutritious food (2008: 12). The economic crisis that commenced in 2008 posed further threats by way of lowering the real wages of the poor and leading to loss of employment altogether. It also threatened the funds available for food which is necessary to avoid malnutrition and starvation (2008: 13). This report also documents increasing levels of social and political unrest in many countries since the beginning of 2007, even in those countries which traditionally have 'high governance performance' (2008: 14). It argues that the variability in food prices, combined with a lack of access to finance and difficulties faced by farmers in financing debts, has made it difficult for small farmers to make new investments (2008: 15). Problems of food shortage will only be exacerbated by the appropriation by international investors of farm lands in many developing world countries. A report published in 2009 documented the large scale land acquisition by foreign private investors in Africa, Latin America, Central Asia, and Southeast Asia (Cotula, et al., 2009). Often underpinned by financial and other support of foreign governments, such acquisition is 'driven' by food security concerns, along with investment in biofuels and increasingly attractive returns in agricultural production (2009: 4–5).

Whereas the biosciences and biotechnologies are often perceived to play a role in alleviating some of the problems faced by the developing nations, such as treating diseases caused by water-borne and insect-borne diseases, many if not most of the problems faced by people in the developing world require attention at the *political and economic levels*. This includes stamping out widespread corruption endemic in many countries, transforming ineffective or corrupt political systems, rethinking trade policies that undermine economic autonomy and reinforce dependence on rich countries for energy, food, and other resources, and investing in local economies to help establish self sufficiency and provide the basis for effective education, health, and welfare systems. There needs to be a more vigorous debate within the rich West about whether investment in many of the high-cost, resource intensive biosciences and biotechnologies is the way of the future. Governments in many countries seem to take it for granted that the bio-economy will provide the basis for the improved health and wellbeing of their populations. However, this is often based on little concrete evidence and little consideration of the impacts on poor people, most of whom are located in the developing world and whose fundamental needs—for sustainable access to nutritious food, clean water, adequate housing, energy security, political stability, and personal safety—currently are not being met. Difficult questions need to be asked about who ultimately benefits and who loses from investment decisions in the biosciences and biotechnologies, and what social values and practices are reinforced by those decisions. Such questions have in the main been marginalised in the wave of enthusiasm that has accompanied emergent or promised new technologies in recent years. If such questions are to be brought to the centre of analysis, it is important that current theoretical perspectives and assumptions are open to scrutiny. In particular, the Northern- and metropolitan-centred focus of theory needs to be recognised, and greater attention needs to be given to the perspectives of peoples in the Southern and non-metropolitan regions. Writing about Sociology's bias towards the Northern metropolitan regions, Connell (2007) highlights how ideas about progress and social evolution have profoundly shaped Western theory, often serving to justify global level inequalities, patterns of dependency, and exploitation. Mainstream bioethics suffers from a similar knowledge bias in its frameworks, reasoning, and repertoire of strategies, which needs to be acknowledged and the implications made evident.

Greater Sensitivity to Global Politics and History

Finally, a critical perspective on the biosciences and biotechnologies will need to pay greater cognisance to questions of global politics and history. Insofar as bioethicists have shown any interest in history, this has been mostly oriented to creating foundation myths rather than employing history to challenge current ways of knowing and acting. Reference has been made in Chapter 1 to the role of bioethics' 'origin stories' in legitimising bioethics' knowledge and practices. Histories can serve another, more radical purpose, however. Following the genealogical approach of the French philosopher and historian, Michel Foucault, histories can help challenge the apparent self-evidence of current ways of knowing and acting by showing that things could be otherwise; hence their designation 'histories of the present' (e.g. Foucault, 1975, 1977). In his histories of medicine, penal institutions, and sexuality, Foucault's histories powerfully underline the contingency of human knowledge and practices. Whereas many histories do this to some extent, the genealogical approach proposed by Foucault links shifting discourses and practices to changes in the mechanisms of power. A shift in medical practice towards a greater understanding of the internal workings of the body and processes of disease, for example, can be seen as an aspect of a more general shift in the workings of power, focussed on the individual body and involving new forms of specialist expertise (Foucault, 1975). Critical scholarship in the field of the biomedicines and biotechnologies could benefit greatly from such an approach, in showing the variability and contingency of the practices and dilemmas that provide the focus for 'ethical' deliberation. As Foucault himself emphasised, 'ethics' itself is a constructed, and hence variable, category, with his later work oriented to developing an ethics of the self that breaks with the normalising tendencies associated with broad moral prescriptions of contemporary societies (Foucault, 1985, 1986).

Whereas Foucault's work is invaluable in rethinking 'ethics' as applied to the biosciences and biotechnologies, it is important when utilising his ideas not to lose sight of the significance of the politico-economic dimensions of science and technology. Mainstream bioethics has been largely blind to the workings of political economy, which is hardly surprising given its primary focus on the dilemmas arising from discrete research and clinical settings, its dependence on principlism, and its over-reliance on moral and analytic philosophy. Questions of political economy were side-lined as the ascendance of neo-liberalism gained pace from the mid-1970s onwards. The view that market rationality should rule in economic decisions and personal life left little space for critique and radical intervention. Issues of health care provision, educational attainment, and personal consumption of goods and services have been seen as largely matters for individual determination and 'choice' in the 'free market'. Neo-liberalism appeared to be so self-evidently the 'correct' course for economic activity and social action that it has proved difficult to dislodge. Universities, no less than other institutions, have been infected by neoliberal rationality, with virtually all fields of knowledge being influenced to varying degrees. In this context, the social sciences and humanities have had to justify their role and show their economic utility. The audit culture, which focuses on quick, measurable 'outputs' and assesses individuals' worth according to their contributions to institutional grant income, has had a profoundly distorting influence on academic work and on how scholars think about their role and the scope and means for change. Critical work has been largely undervalued. The collapse of the global economic system in 2008–2009, however, generated much debate about the failures of neo-liberalism and the role of the state in regulating economic and financial systems. This event, combined with growing concerns about climate change, has brought into question the future sustainability of capitalism.

The fate of the biosciences and biotechnologies has been tied strongly to the fate of global, consumer capitalism. The direction and level of investment in biotechnology and the pharmaceutical industries has shifted over time according to economic factors and related political conditions. This was brought sharply into focus with the financial crisis of 2008–2009. Some of the supposedly greatest 'breakthroughs' in the biosciences in recent years, in so-called gene-mapping, occurred with considerable private and public investment during a period of economic boom in many countries. The Human Genome Project (HGP) was the most visible aspect of this investment. In 2008–2009, however, biotech shares collapsed with R&D on many new purportedly promising innovations following in the wake of the HGP and other genetic innovations shelved (see Chapter 2). Even before this event, the impacts of economic decisions and the rise and fall of the share market on the fortunes of the biotech and pharmaceutical sectors was clearly evident. A truly critical perspective on the biosciences and biotechnologies would pay cognisance to the workings of political economy and seek to understand how this shapes and is likely in the future to shape innovations. One of my goals in writing this book has been to highlight the neglected politico-economic dimension of not only the biosciences and biotechnologies but of bioethics itself. This book is but a beginning of a much needed rethinking of this field. I hope that it will generate debate and invite critical comment, which can only serve to help change entrenched thinking and practice which is urgently needed to meet the challenges of the future.

References

Abraham, J. (1995) *Science, Politics and the Pharmaceutical Industry: Controversy and Bias in Drug Regulation*. UCL Press: London.

Allen, G. (1996) 'Science misapplied: The eugenic age revisited', *Technology Review*, 99, 6 (August/September), 23–31.

American Board of Genetic Counseling, Inc. (2008) *ABGC Certified Genetic Counselors*. *Brochure*. http://www.abgc.net/CMFiles/ABGC_Brochure51OQS-10242008–3600.pdf. Accessed 28 May 2010.

Anderson, A., Allan, S., Petersen, A. & Wilkinson, C. (2005) 'The framing of nanotechnologies in the British newspaper press', *Science Communication*, 27, 2: 1–21.

Anderson, A., Petersen, A., Wilkinson, C. and Allan, S. (2009) *Nanotechnology, Risk and Communication*. Palgrave Macmillan: Houndmills.

Anker, S. and Nelkin, D. (2004) *The Molecular Gaze: Art in the Genetic Age*. Cold Spring Harbor Laboratory Press: Cold Spring Harbour, NY.

Aramesh, K. and Dabbagh, S. (2007) 'An Islamic view to stem cell research and cloning: Iran's experience', *The American Journal of Bioethics*, 7, 2: 62–75.

Armstrong, D. (1995) 'The rise of surveillance medicine', *Sociology of Health and Illness*, 17, 3: 393–404.

Asai, A., Yamamoto, W. and Fukui T. (1997) 'What ethical dilemmas are Japanese physicians faced with?' *Eubios Journal of Asian and International Bioethics*, 7: 162.

Associated Press (2009) 'Obama reverses Bush-era stem cell policy', *msmbc.com*, March 9. http://www.msnbc.msn.com/id/29586269/. Accessed 15 February 2010.

Association for Clinical Biochemistry (2006) *Modernising Pathology: Building a Service Responsive to Patients*. Submission from the Association for Clinical Biochemistry to The Pathology Service Review Panel, January 2006. Executive Summary. http://www.acb.org.uk/docs/ACBCarterSubmission.pdf. Accessed 16 April 2010.

Australian Council of Trade Unions (2009) *Nanotechnology—Why Unions are Concerned*. Fact Sheet April 2009. http://www.actu.org.au/Images/Dynamic/attachments/6494/actu_factsheet_ohs_-nanotech_090409.pdf. Accessed 6 July 2010.

Australian Health Care Associates (n.d.) *Genetic Services Strategy for Victoria 2005–2009*. http://www.health.vic.gov.au/ Accessed 28 October 2010.

Australian Medical Association (2008). *Direct-to-Consumer Advertising*, February 2008. http://www.ama.com.au/node/2931. Accessed 16 April 2010.

Australian Office of Nanotechnology (2009) *National Nanotechnology Strategy Annual Report 2008–09*. Australian Office of Nanotechnology: Canberra.

http://www.innovation.gov.au/Industry/Nanotechnology/Documents/annual-report0910.pdf. Accessed 10 July 2010.

Bachrach, P. and Baratz, M. S. (1963) 'Decisions and nondecisions: an analytic framework', *The American Political Science Review*, 57, 3: 632–642.

Bailey, R. (1996) 'Prenatal testing and the prevention of impairment: a women's right to choose?', in J. Morris (ed) *Encounters With Strangers: Feminism and Disability*. The Women's Press Ltd: London.

Barr, M. (2006) ' "I'm not really read up on genetics": biobanks and the social context of informed consent', *BioSocieties*, 1: 252–262.

Baylis, F. (2008) 'Animal eggs for stem cell research: a path not worth taking', *The American Journal of Bioethics*, 8, 12: 18–32.

Beauchamp, R. (2009) 'Special offer for "genome scan" customers willing to participate in research', *BIONEWS*, no. 518, 26 July. http://www.bionews.org.uk/page_45620.asp?iruid=294, Accessed 31 March 2010.

Beauchamp, T. L. and Childress J. F. (2001) *Principles of Biomedical Ethics*. 5th Edition. Oxford University Press: New York.

Biotechnology Industry Organisation (2009) 'Principles on personalized medicine'. http://www.bio.org/healthcare/personalized/BIO_PM_Principles_FINAL.pdf. Accessed 17 April 2009.

Bird, K. (2010) 'France's public nano-debate suffers from continued protests', *Cosmetics design.com*, 24 February. http://www.cosmeticsdesign.com/Formulation-Science/France-s-public-nano-debate-suffers-from-continued-protests. Accessed 6 July 2010.

Bock, D. (2010) *The Politics of Happiness: What Government Can Learn From the New Research on Well-Being*. Princeton University Press: Princeton and Oxford.

Borry, P., Schotsmans, P. and Dierickx, K. (2006) 'Author, contributor or just a signer?: a quantitative analysis of authorship trends in the field of bioethics', *Bioethics*, 20, 4: 213–220.

Boseley, S. (2008) 'Adverse drug reactions cost NHS £2 billion', *The Guardian*, 3 April. http://www.guardian.co.uk/society/2008/apr/03/nhs.drugsandalcohol. Accessed 14 August 2009.

Bosk, C. L. (1992) *All God's Mistakes: Genetic Counseling in a Pediatric Hospital*. The University of Chicago Press: Chicago and London.

Bourdieu, P. (1986) *Distinction: A Social Critique of the Judgement of Taste*. Routledge and Kegan Paul: London.

Boyes, M. (1999) 'Whose DNA? Genetic surveillance, ownership of information and newborn screening', *New Genetics and Society*, 18, 2/3, 145–155.

Brice, P. (2010) 'Making the most of genomics for health', *BIONEWS*, 559, 24 May. http://www.bionews.org.uk/page_61547.asp?dinfo=XdPLnoBGA55enlspQ0TDjuBO. Accessed 25 May 2010.

Brock, S. C. (1995) 'Narrative and medical genetics: on ethics and therapeutics', *Qualitative Health Research*, 5: 150–168.

Brody, H. (2009) *The Future of Bioethics*. Oxford University Press: Oxford and New York.

Bubela, T. M., Caulfield T. A. (2004) 'Do the print media "hype" genetic research?: A comparison of newspaper stories and peer-reviewed research papers', *Canadian Medical Association Journal*, 170, 9: 1399–1407.

Bunton, R. and Petersen, A. (eds) (2005) *Genetic Governance: Health, Risk and Ethics in the Biotech Era*. Routledge: London and New York.

Cambridge Advanced Learner's Dictionary (2010) 'Imperialism'. http://dictionary.cambridge.org/define.asp?key=39341&dict=CALD&topic=colonisation-and-self-government. Accessed 19 March 2010.

Capps, B. (2008) 'Authoritative regulation and the stem cell debate', *Bioethics*, 22, 1: 43–55.

Caulfield, T. (2004) 'The commercialisation of medical and scientific reporting', *PloS Medicine*, 1, 3: 178–179.

Caulfield, T. and Brownsword, R. (2006) 'Human dignity: a guide to policy making in the biotechnology era?', *Nature Reviews Genetics*, 7: 72–76.

Caulfield, T. and Bubela, T. (2007) 'Why a criminal ban?: analysing the arguments against somatic cell nuclear transfer in the Canadian Parliamentary Debate', *American Journal of Bioethics*, 7, 2: 51–61.

Caulfield, T. and Chapman, A. (2005) 'Human dignity as a criterion for science', *PLoS*, 2, 8 (August): 0736–0738. www.plosmedicine.org. Accessed 8 March 2010.

Chadwick, R. (1993) 'What counts as success in genetic counselling?', *Journal of Medical Ethics*, 19: 43–46.

Chadwick, R., Levitt, M. and Shickle, D. (eds) (1997) *The Right to Know and the Right not to Know*. Ashgate: Aldershot.

Chapman, A. and Hiskes, A. L. (2008) *The American Journal of Bioethics*, 8, 12: 44–46.

Crichton, M. (2002) *Prey*. Harper Collins: New York.

Clapton, J. (2003) 'Tragedy and catastrophe: contentious discourses of ethics and disability', *Journal of Intellectual Disability Research*, 47, 7: 540–547.

Clarke, A. (1990) 'Genetics, ethics and audit', *The Lancet*, 335: 1145–1147.

Clarke, A. (1991) 'Is non-directive counselling possible?', *The Lancet*, 338: 998–1001.

Connell, R. (2007) *Southern Theory: The Global Dynamics of Knowledge in Social Science*. Allen and Unwin: Crows Nest, NSW.

Conrad, P. (1999) 'Uses of expertise: Sources, quotes, and voice in the reporting of genetics in the news', *Public Understanding of Science*, 8: 285–302.

Corrigan, O. (2003) 'Empty ethics: the problem with informed consent', *Sociology of Health and Illness*, 25, 3: 768–792.

Corrigan, O., McMillan, J., Liddell, K., Richards, M. and Weijer, C. (eds) (2009) *The Limits of Consent: A Socio-Ethical Approach to Human Subject Research in Medicine*. Oxford University Press: Oxford.

Corrigan, O. and Petersen, A. (2008) 'UK Biobank: bioethics as a technology of governance', in H. Gottweis and A. Petersen (eds) *Biobanks: Governance in Comparative Perspective*. Routledge: London and New York.

Corrigan, O. and Tutton, R (eds) (2004) *Genetic Databases: Socio-Ethical Issues in the Collection and Use of DNA*. Routledge: London.

Cotula, L., Vermeulen, S., Leonard, R. and Keeley, J. (2009) *Land grab or development opportunity?: Agricultural investment and international land deals in Africa*. IIED/FAO/IFAD: London and Rome.

Cunningham-Burley, S. (2006) 'Public knowledge and public trust', *Community Genetics*, 9: 204–210.

Davies, S. M. (2006) 'Pharmacogenetics, pharmacogenomics and personalized medicine: are we there yet?', *Hematology*, 1: 111–117.

De Vries, R. (2003) 'How can we help?: From "sociology in" to "sociology of" Bioethics', *The Journal of Law, Medicine and Ethics*, 32: 279–292.

De Vries, R. (2007) '(Bio)ethics and evidence: from collaboration to co-operation', in C. Gastmans, K. Dierickx, H. Nys and P. Schotsmans (eds) *New Pathways for European Bioethics*. Intersentia: Antwerpen, Oxford.

De Vries, R., Turner, L., Orfali K. and Bosk, C. (eds) (2007) *The View From Here: Bioethics and the Social Sciences*. Blackwell Publishing: Malden, MA.

Dennis, K. and Urry, J. (2009) *After the Car*. Polity: Cambridge.

Department of Health (2003) *Our Inheritance, Our Future: Realising the Potential of Genetics in the NHS*. HMSO: Norwich.

Djelic, M.-L. and Quack, S. (2007) 'Overcoming path dependency: path generation in open systems', *Theory and Society*, 36: 161–186.

Dobkin, B. H., Curt, A. and Guest, J. (2006) 'Neuro-rehabilitation, *Neural Repair*, 20: 5–13.

Donchin, A. (2001) 'Understanding autonomy relationally: toward a reconfiguration of bioethical principles', *Journal of Medicine and Philosophy*, 26, 4: 365–386.

Donchin, A. (2008) 'Remembering FAB's past, anticipating our future', The *International Journal of Feminist Approaches to Bioethics*, 1, 1: 145–160.

Donohue, J. M., Cevasco, M. and Rosenthal, M. B. (2007) 'A decade of direct-to-consumer advertising of prescription drugs', *The New England Journal of Medicine*, 357, 7: 673–681.

Döring, O. (2003) 'Chinese researchers promote biomedical regulations: what are the motives of the biopolitical dawn in China and where are they heading?', *Kennedy Institute of Ethics Journal*, 14, 1: 39–46.

Doward, J. (2004) 'Tobacco giant funds "bad gene" hunt', *The Observer*, guardian.co.uk.   http://www.guardian.co.uk/society/2004/may/30/health.smoking. Accessed 16 July 2010.

Duster, T. (1990) *Backdoor to Eugenics*. Routledge: New York.

Edwards, I. R. and Aronson, J. K. (2000) 'Adverse drug reactions: definitions, diagnosis and management', *The Lancet*, 356, 9237: 1205–1284.

Eensaar, R. (2008) 'Estonia: ups and downs of a biobank project', in H. Gottweis and A. Petersen (eds) *Biobanks: Governance in Comparative Perspective*. Routledge: London and New York.

Elger, B., Biller-Andorno, N., Mauron, A. and Capron, A. (2008) *Ethical Issues in Governing Biobanks: Global Perspectives*. Ashgate: Hampshire and Birlington.

Elliott, C. (2003) *Better Than Well: American Medicine Meets the American Dream*. WW Norton: New York.

Elliott, K. (2007) 'An ironic *deductio* for a "pro-life" argument: Hurlbut's proposal for stem cell research', *Bioethics*, 21, 2: 98–110.

ETC Group (2003) *No Small Matter 2: The Case for a Global Moratorium*. Occasional Paper Series. April 2003. http://www.etcgroup.org/upload/publication/165/01/occ.paper_nanosafety.pdf. Accessed 6 July 2010.

European Commission (2010) *Communicating Nanotechnology: Why, To Whom, Saying What and How?* European Commission: Luxembourg. http://ec.europa.eu/research/industrial_technologies/pdf/communicating-nanotechnology_en.pdf. Accessed 2 June 2010.

Evans, J. (2002) *Playing God?: Human Genetic Engineering and the Rationalization of Public Bioethical Debate*. The University of Chicago Press: Chicago and London.

Farmer, P. (2003) *Pathologies of Power: Health, Human Rights and the New War on the Poor*. University of California Press: Berkeley.

Farmer, P. and Campos, N. G. (2004) 'New malaise: bioethics and human rights in the global era', *Journal of Law, Medicine and Ethics*, 32: 243–251.

Fatimathas, L. (2010) 'First synthetic cell created in a laboratory', *BIONEWS 559*, 24 May 2010. http://www.bionews.org.uk/page_61470.asp. Accessed 6 August 2010.

Feero, W. G., Guttmacher, A. E. and Collins, F. S. (2008) 'The genome gets personal—almost', *JAMA*, 299, 11: 1351–1352.

Fortin, S. and Knoppers, B. M. (2009) 'Secondary uses of personal data for population research', *Genomics, Society and Policy*, 5, 1: 80–99.

Foucault, M. (1975) *The Birth of the Clinic: An Archaeology of Medical Perception*. Vintage Books: New York.

Foucault, M. (1977) *Discipline and Punish: The Birth of the Prison*. Penguin: Harmondsworth, Middlesex and New York.

Foucault, M. (1985) *The Use of Pleasure*. Trans. R. Hurley. Penguin: Harmondsworth.

Foucault, M. (1986) *The Care of the Self*. Trans. R. Hurley. Penguin: Harmondsworth.

Fox, R. and Swazey, J. P. (1978) *The Courage to Fail: A Social View of Organ Transplants and Dialysis*. University of Chicago Press: Chicago.

Fox, R. and Swazey, J. P. (1992) *Spare Parts: Organ Replacement in American Society*. Oxford University Press: New York.

Fox, R. and Swazey, J. P. (2008) *Observing Bioethics*, Oxford University Press: Oxford.

Gavelin, K., Wilson, R. and Doubleday, R. (2007) *Democratic Technologies?: The Final Report of the Nanotechnology Engagement Group*. Involve: London.

GeneWatch UK (2009) *Biosciences for Life?: Appendix A: The History of UK Biobank, Electronic Medical Records in the NHS, and the Proposal for Data-Sharing Without Consent*. January 2009. GeneWatch UK: Buxton, Derbyshire. http://genewatch.org/uploads/f03c6d66a9b354535738483c1c3d49e4/UK_Biobank_fin_1.pdf. Accessed 19 July 2010.

GeneWatch UK (2010a) *History of the Human Genome*. Briefing paper. June. http://www.genewatch.org/uploads/f03c6d66a9b354535738483c1c3d49e4/HGPhistory_2.pdf. Accessed 16 July 2010.

GeneWatch UK (2010b) *Bioscience for Life?: Who Decides What Research is Done in Health and Agriculture?* GeneWatch UK: Buxton, Derbyshire.

Ghosh, P. (2010) 'Journal stem cell work "blocked" ', *BBC News*, 2 February. http://news.bbc.co.uk/2/hi/science/nature/8490291.stm. Accessed 12 February 2010.

Gibbon, S. and Novas, C. (2008) *Biosocialities, Genetics and the Social Sciences*. Routledge: Abingdon and New York.

Gibson, I. (2002) Petition on 'Biobank': House of Commons Hansard Debates, 3 July 2002. http://www.parliament.the-stationery-office.co.uk/pa/cm200102/cmhansrd/vo020703/debtext/20703–43.htm. Accessed 13 August 2010.

Gieryn, T. F. (1999) *Cultural Boundaries of Science: Credibility on the Line*. Chicago University Press: Chicago.

Ginsborg, P. (2008) *Democracy: Crisis and Renewal*. Profile Books: London.

Goddard, K. A. B., Robitaille, J., Dowling, N. F., Parrado, A. R., Fishman, J., Bradley, L. A., Moore, C. A. and Khoury, M. J. (2009) 'Health-related direct-to-consumer genetic tests: a public health assessment and analysis of practices related to Internet-based tests for risk of Thrombosis', *Public Health Genomics*, 12: 92–104.

Goering, S. (2008) ' "You say you're happy . . . but": contested quality of life judgements in bioethics and disability studies', *Bioethical Inquiry*, 5: 125–135.

Gottweis, H. and Kim, B. (2010) 'Explaining Hwang-Gate: South Korean identity politics between bionationalism and globalization', *Science, Technology and Human Values*, 35, 4: 501–524.

Gottweis, H. and Petersen, A. (eds) (2008) *Biobanks: Governance in Comparative Perspective*. Routledge: London and New York.

Gottweis, H., Salter, B. and Waldby, C. (2009) *The Global Politics of Human Embryonic Stem Cell Science: Regenerative Medicine in Transition*. Palgrave Macmillan: New York.

Green, R. M. (2002) 'Benefiting from "evil": an incipient moral problem in human stem cell research', *Bioethics*, 16, 2: 544–556.

Green, S. K. (2007) 'Is Canada's stem cell legislation unwittingly discriminatory?', *The American Journal of Bioethics*, 7, 8: 50–52.

Guilleman, J. H. and Holmstrom, L. L. (1986) *Mixed Blessings: Intensive Care for Newborns*. Oxford University Press: New York.

Guston, D. and Sarewitz, D. (2002) 'Real-time technology assessment', *Technology in Society*, 24: 93–109.

Hallowell, N. (1999) 'Doing the right thing: genetic risk and responsibility', *Sociology of Health and Illness*, 21: 597–621.

Hallowell, N. (2009) 'Consent to genetic testing: a family affair?', in O. Corrigan, J. McMillan, K. Liddell, M. Richards and C. Weijer (eds) *The Limits of Consent: A Socio-Ethical Approach to Human Subject Research in Medicine.* Oxford University Press: Oxford.

Hallowell, N., Foster, C., Eeles, R., Ardern-Jones, A., Murday, V. and Watson, M. (2003) 'Balancing autonomy and responsibility: the ethics of generating and disclosing genetic information', *Journal of Medical Ethics*, 29: 74–83.

Hannaford, I. (1996) *Race: The History of an Idea in the West.* The Woodrow Wilson Centre Press: Washington, and the John Hopkins University Press: Baltimore.

Hansen, M. J. (1999) 'Biotechnology and commodification within health care', *Journal of Medicine and Philosophy*, 24, 3, 267–287.

Hård, M. and Jameson, A. (2005) *Hubris and Hybrids: A Cultural History of Technology and Science.* New York: Routledge.

Harris, R. (1998) 'Genetic counselling and testing in Europe', *Journal of the Royal College of Physicians of London,* 32: 335–338.

Harrison, D. (2009) 'Unions call for action to oversee nanotechnology', *the age.com.au*, April 14. http://www.theage.com.au/national/unions-call-for-action-to-oversee-nanotechnology-20090413-a4ts.html. Accessed 4 June 2010.

Hashiloni-Dolev, Y. and Raz, A. E. (2010) 'Between social hypocrisy and social responsibility: professional views of eugenics, disability and repro-genetics in Germany and Israel', *New Genetics and Society*, 29, 1: 87–102.

Hedgecoe, A. M. (2004a) 'Critical bioethics: beyond the social science critique of applied ethics', *Bioethics*, 18, 2: 120–143.

Hedgecoe, A. M. (2004b) *The Politics of Personalised Medicine: Pharmacogenetics in the Clinic.* Cambridge University Press: Cambridge.

Hedgecoe, A. M. (2010) 'Bioethics and the reinforcement of socio-technical expectations', *Social Studies of Science*, 40, 2: 163–186.

Henderson, M. (2010) 'Genetic testing to get flawless babies', *The Australian*, February 9, p. 3.

Henen, L., Sauter, A. and Van Den Cruyce, E. (2010) 'Direct to consumer genetic testing: insights from an internet scan', *New Genetics and Society*, 29, 2: 167–186.

Hennig, W. (2006) 'Bioethics in China', *European Molecular Biology Reports*, 7, 9: 850–854.

Ho, A. (2008) 'The individualist model of autonomy and the challenge of disability', *Bioethical Inquiry*, 5: 193–207.

Hollands, P. and McCauley, C. (2009) 'Private cord blood banking: current use and clinical future', *Stem Cell Reviews and Reports*, 5: 195–203.

Holm, S. and Madsen, S. (2009) 'Informed consent in medical research—a procedure stretched beyond breaking point?', in O. Corrigan, J. McMillan, K. Liddell, M. Richards and C. Weijer (eds) *The Limits of Consent: A Socio-Ethical Approach to Human Subject Research in Medicine.* Oxford University Press: Oxford.

Holtzman, N. A. and Marteau,T. M. (2000) Will genetics revolutionize medicine?, *New England Journal of Medicine*, 34, 2: 141–144.

Hood, L. (2009) 'A doctor's vision of the future of medicine', *Newsweek*, July 6/ July 13: 50.

House of Lords Select Committee on Science and Technology (2000) *Science and Society: Third Report.* London: HMSO. http://www.publications.parliament. uk/pa/ld199900/ldselect/ldsctech/38/3801.htm. Accessed 12 August 2010.

House of Commons Select Committee on Science and Technology (2003) *Third Report* (March). http://www.publications.parliament.uk/pa/ld200304/ldselect/ldsctech/ldsctech.htm. Accessed 12 August 2010.

House of Commons Select Committee on Science and Technology (2004) *Too Little Too Late? Government Investment in Nanotechnology. Fifth Report of Session 2003–04. Volume 1.* The Stationery Office: London.

House of Lords Science and Technology Committee (2009) *Genomic Medicine.* Volume 1: Report. The Stationery Office: London. http://www.publications.parliament.uk/pa/ld200809/ldselect/ldsctech/107/107i.pdf. Accessed 25 May 2010.

Human Genetics Commission (2007) *More Genes Direct: A Report on the Availability, Marketing and Regulation of Genetic Tests Supplied Directly to the Public.* December 2007. Human Genetics Commission: London.

Human Genetics Commission (2010) *A Common Framework of Principles for Direct-to-Consumer Genetic Testing Services.* http://www.hgc.gov.uk/UploadDocs/DocPub/Document/HGC%20Principles%20for%20DTC%20genetic%20tests%20-%20final.pdf. Accessed 10 August 2010.

Human Variome Project Bulletin (2010) August–February 2010. N.17. http://www.humanvariomeproject.org/images/stories/bulletins/hvp_bulletin_17.pdf. Accessed 9 April 2010.

Humber, J. M. and Almeder, R. F. (eds) (1998) *Human Cloning.* Humana Press: Totowa, NJ.

Hutchings, J. (2009) 'A transformative Māori approach to bioethics', in Te Mata o te Tau (ed) *Matariki.* Te Mata o te Tau Monograph: Wellington, New Zealand.

Huxley, A. (1955; orig. 1932) *Brave New World.* Penguin Books: Harmondsworth.

Hyder, N. (2010) 'deCODE is back', *BIONEWS*, 25 January. http://www.bionews.org.uk/page_53705.asp?iruid=294. Accessed 5 February 2010.

Ikediobi, O.N. (2009) 'Personalized medicine: are we there yet?', *The Pharmacogenomics Journal*, 9: 85.

Irwin, A. and Michael, M. (2003) *Science, Social Theory and Public Knowledge.* Open University Press: Maidenhead and Philadelphia.

Isas, R. M. (2009) 'Policy inoperability in stem cell research: demystifying harmonization', *Stem Cell Reviews and Reports*, 5: 108–115.

Ishizu, S., Sekiya, M., Ishibashi, K., Negami, Y. and Ata, M. (2008) 'Towards the responsible innovation with nanotechnology in Japan: our scope', *Journal of Nanoparticle Research*, 10, 2: 229–254.

Jackson, T. (2009) *Prosperity Without Growth: Economics for a Finite Planet.* Earthscan: London.

Jasanoff, S. (2005). *Designs on Nature: Science and Democracy in Europe and the United States.* Princeton University Press: Princeton, NJ, and Oxford.

Jensen, E. (2008) 'The Dao of human cloning: utopian/dystopian hype in the British press and popular films', *Public Understanding of Science*, 17, 2: 123–143.

Jones, B. (2009) 'Recession threatens UK stem cell progress', *BIONEWS*, 503, 7–14 April. Progress Education Trust: London.

Jonsen, A. R. (1998) *The Birth of Bioethics.* Oxford University Press: New York.

Jung, K. W. and Hyun, I. (2006) 'Oocyte and somatic cell procurement for stem cell research: The South Korean experience', *The American Journal of Bioethics*, 6, 1: W19–W22.

Kaelin, L. (2009) 'Contextualizing bioethics: the UNESCO declaration on bioethics and human rights and observations on Filipino bioethics', *Eubios Journal of Asian and International Bioethics*, 19: 42–47.

Kass, L. R. (2003) *Beyond Therapy: Biotechnology and the Pursuit of Happiness.* A Report by the President's Council on Bioethics. ReganBooks: New York.

Katz, E., Lovel, R., Mee, W. and Solomon, F. (2005) *Citizens' Panel on Nanotechnology. Report to Participants*. CSIRO Minerals: Clayton.

Katz, E., Solomon, F., Mee, W. and Lovel, R. (2009) 'Evolving scientific research governance in Australia: a case study of engaging interested publics in nanotechnology research', *Public Understanding of Science*, 18, 5: 531–545.

Kemp, E. (2009) 'Open letter to Senior Editors of peer-review journals publishing in the field of stem cell biology', *EuroStemCell*, 10 July. http://eurostemcell.org/commentanalysis/peer-review. Accessed 12 February 2010.

Kenen, R. H. (1986) 'Growing pains of a new health care field genetic counselling in Australia and the United States', *Australian Journal of Social Issues*, 21:172–182.

Kian, C. T. S. and Leng, T. S. (2005) 'The Singapore approach to human stem cell research, therapeutic and reproductive cloning', *Bioethics*, 19, 3: 290–303.

Kimmelman, J. (2007) 'Towards a global human embryonic stem cell bank: differential termination', *The American Journal of Bioethics*, 7, 8: 52–53.

Kitzinger, J. (2008) 'Questioning hype, rescuing hope?: The Hwang stem cell scandal and the reassertion of hopeful horizons', *Science as Culture*, 17, 4: 417–434.

Kjølberg, K. and Wickson, F. (2007) 'Social and ethical interactions with nano: mapping the early literature', *NanoEthics*, 1: 89–104.

Knoppers, B. M. (2005) 'Biobanking: international norms', *Regulation of Biobanks*, Spring: 7–14.

Kolata, G. (1997) *Clone: The Road to Dolly and the Path Ahead*. Allen Lane: London.

Komesaroff, P. (2008) *Experiments in Love and Death: Medicine, Postmodernism, Microethics and the Body*. Melbourne University Press: Melbourne.

Kuhse, H. and Singer, P. (2006) *Bioethics: An Anthology*. Blackwell Publishing: Malden, MA and Oxford.

Kurath, M. and Gisler, P. (2009) 'Informing, involving or engaging?: science communication, in the ages of atom-, bio- and nanotechnology', *Public Understanding of Science*, 18, 5: 559–573.

Langreth, R. (2008) 'Biotech blues', *Forbes Magazine*, 12 January 2009. http://www.forbes.com/forbes/2009/0112/090.html. Accessed 10 April 2009.

Latour, B. and Woolgar, S. (1986) *Laboratory Life*. Second Edition. Princeton University Press: Princeton, NJ.

Lau, D., Ogbogu, U., Taylor, B., Stafinski, T., Menon, D. and Caulfield, T. (2008) 'Stem cell clinics online: the direct-to-consumer portrayal of stem cell medicine', *Cell Stem Cell*, 3 December 4. http://www.newscientist.com/article/dn14137-new-task-force-to-tackle-stemcell-tourism.html. Accessed 4 September 2009.

Law, J. (2005) *Big Pharma: How the World's Biggest Drug Companies Control Illness*. Constable: London.

Lederbogen, U. and Trebbe, J. (2003) 'Promoting science on the web: public relations for scientific organizations—results of a content analysis', *Science Communication*, 24: 333–352.

Lenk, C. and Biller-Andorno, N. (2006) 'Nanomedicine—emerging or re-emerging ethical issues?: A discussion of four ethical themes', *Medicine, Health Care and Philosophy*, 10:173–184.

Leonard, C. O., Chase, G. A. and Childs, B. (1972) 'Genetic counseling: a consumer's view', *The New England Journal of Medicine*, 287: 433–439.

Lewis, R. and Zhdanov, R. I. (2009) 'Centenarians as stem cell donors', *The American Journal of Bioethics*, 9, 11: 1–3.

Little, K. (2002) *Making History: Bioethics, Culture and the History of Moral Ideas*. Master's Thesis, submitted to School of Philosophy and Bioethics, Monash University.

Little, M. O. (1996) 'Why a feminist approach to bioethics?', *Kennedy Institute of Ethics Journal*, 6, 1: 1–18.

Liu, H.-E. (2005) 'Regulation and the social response to establishing a large-scale biobank in Taiwan', Paper presented at the annual meeting of the Law and Society, J.W. Marriott Resort, Las Vegas, NV.

López, J. (2004) 'How sociology can save bioethics . . . maybe', *Sociology of Health and Illness*, 26, 7: 875–896.

Lott, J. P. and Savulescu, J. (2007) 'Towards a global human embryonic stem cell bank', *The American Journal of Bioethics*, 7, 8: 37–44.

Maasen, S. and Weingart, P. (2000) *Metaphors and the Dynamics of Knowledge*. Routledge: London and New York.

Macklin, R. (1999) *Against Relativism: Cultural Diversity and the Search for Ethical Universals*. Oxford University Press: New York.

McNamara, B. and Petersen, A. (2008) 'Framing consent: the politics of "engagement" in an Australian biobank project', in H. Gottweis and A. Petersen (eds) *Biobanks: Governance in Comparative Perspective*. Routledge: London and New York.

Macnamara, J. (2009) 'Journalism and PR: beyond myths and stereotypes to transparency and management in the public interest', Unpublished manuscript.

Maienschein, J. (2003) *Whose View of Life?: Embryos, Cloning and Stem Cells*. Harvard University Press: Cambridge, MA.

Maienschein, J. (2009) 'Regenerative medicine in historical context', *Medicine Studies*, 1, 1: 33–40.

Maienschein, J., Sunderland, M., Ankeny, R. A. and Robert, J. S. (2008) 'The ethos and ethics of translational research', *The American Journal of Bioethics*, 8, 3: 43–51.

Malsch, I. and Hvidtfelt-Nielsen, K. (2009) *Individual and Collective Responsibility for Nanotechnology*. First Annual Report on Ethical and Social Aspects of Nanotechnology. ObservatoryNano. Department of Science Studies: University of Aarhus, Denmark. .http://www.observatorynano.eu/project/filesystem/files/annrep1responsibility1.pdf. Accessed 6 July 2010.

Malsch, I. and Hvidtfelt-Nielsen, K. (2010) *Nanobioethics*. ObservatoryNano 2nd Annual Report on Ethical and Societal Aspects of Nanotechnology, 21 April 2010. http://www.observatorynano.eu/project/catalogue/4NB/. Accessed 2 July 2010.

Mansfield, P. R., Mintzes, B., Richards, D. and Toop, L. (2005) 'Direct to consumer advertising', *British Medical Journal*, 330: 5, 6.

Market Attitude Research Services (2008) *Australian Community Attitudes Held About Nanotechnology—Trends 2005 to 2008*. DIISR: Canberra.

McCleod, C. and Baylis, F. (2007) 'Donating fresh versus frozen embryos to stem cell research: in whose interests?', *Bioethics*, 21, 9: 465–477.

McGee, G. (ed) (1998) *The Human Cloning Debate*. Berkeley Hills Book: Berkeley, CA.

McNamara, B. and Petersen, A. (2008) 'Framing consent: the politics of "engagement" in an Australian biobank project', in H. Gottweis and A. Petersen (eds) *Biobanks: Governance in Comparative Perspective*. Routledge: London and New York.

Medew, J. and Leung, C. C. (2008) 'Infighting clouds stem cell centre's future', *The Age*, September 6. http://www.theage.com.au/national/infighting-clouds-stem-cell-centres-future-20080905–4aq4.html. Accessed 16 September 2009.

Mee, W., Lovel, R., Solomon, F., Kearns, A., Cameron, F. and Turney, T. (2004) *Nanotechnology: The Bendigo Workshop*. DMR-2561. CSIRO Minerals: Clayton South: Victoria.

Miringoff, M.-L. (1991). *The Social Costs of Genetic Welfare.* Rutgers University Press: New Brunswick, NJ.

Morrison, D. R. (2008) 'Making the autonomous client: how genetic counsellors construct autonomous subjects', in B. Katz Rothman, E. M. Armstrong and R. Tiger (eds), *Biological Issues, Sociological Perspectives.* Advances in Medical Sociology, Volume 9, pp. 179–198.

Moynihan, R. and Cassels, A. (2005) *Selling Sickness: How Drug Companies Are Turning Us All Into Patients.* Allen & Unwin: Crows Nest, NSW.

Murphy, T. F. (2008) 'When is an objection to hybrid stem cell research a moral objection?', *The American Journal of Bioethics*, 8, 12: 47–49.

National Institutes of Health (2010) Research Portfolio Online Reporting Tools (RePORT). http://report.nih.gov/rcdc/categories/default.aspx. Accessed 2 August 2010.

National Science and Technology Council (2000) *National Nanotechnology Initiative: The Initiative and Its Implementation Plan.* National Science and Technology Council, Committee on Technology, Subcommittee on Nanoscale Science, Engineering and Technology: Washington, DC. July 2000. http://www.wtec.org/loyola/nano/IWGN.Implementation.Plan/nni.implementation.plan.pdf. Accessed 7 July 2010.

National Science and Technology Council (2007) *The National Nanotechnology Initiative. Strategic Plan 2007.* National Science and Technology Council: Arlington, VA.

Nelkin, D. (1987) *Selling Science: How the Press Covers Science and Technology.* W.H. Freeman: New York.

Nelson, L. J. and Meyer, M. J. (2005) 'Confronting deep moral disagreement: the President's Council on Bioethics, Moral Status, and Human Embryos', *The American Journal of Bioethics*, 5, 6: 33–42.

Newell, C. (2006) 'Disability, bioethics, and rejected knowledge', *Journal of Medicine and Philosophy*, 31: 269–283.

Nisbet, M. C. and Lewenstein, B. (2002) 'Biotechnology and the American media: the policy process and the elite press, 1970 to 1999', *Science Communication*, 23: 359.

Nisbet, M. C. and Scheufele, D. A. (2009) 'What's next for science communication?: promising directions and lingering distractions', *American Journal of Botany*, 96, 10: 1767–1778.

Nussbaum, M. C. and Sunstein, C. R. (eds) (1998) *Clones and Clones.* W.W. Norton and Co: New York and London.

Oakley, J. (ed) (2009) *Bioethics.* Ashgate: Farnham, Surrey.

Oudshoorn, N. and Pinch, T. (eds) (2003) *How Users Matter: The Co-Construction of Users and Technologies.* The MIT Press: Cambridge, MA and London.

Outomuro, D. (2007) 'Moral dilemmas around a global human embryonic stem cell bank', *The American Journal of Bioethics*, 7, 8: 47–48.

Oxfam International (2009) *Suffering the Science: Climate Change, People and Poverty.* 6 July 2009. www.oxfam.org

Pálsson, G. (2007) *Anthropology and the New Genetics.* Cambridge University Press: Cambridge.

Pálsson, G. (2008) 'The rise and fall of a biobank: the case of Iceland', in H. Gottweis and A. Petersen (eds) (2008) *Biobanks: Governance in Comparative Perspective.* Routledge: London and New York.

Pálsson, G. and Hardardóttir, K. E. (2002) 'For whom the cell tolls: debates about biomedicine', *Current Anthropology*, 43, 2: 271–301.

Patra, P. K. and Sleeboom-Faulkner, M. (2009) 'Informed consent in genetic research and biobanking in India: some common impediments', *Genomics, Society and Policy*, 5, 1: 100–113.

Paul, D. B. (1998) *The Politics of Heredity: Essays on Eugenics, Biomedicine, and the Nature-Nurture Debate*. State University of New York Press: New York.

Pavlopoulos, M., Grinbaum, A. and Bontems,V. (2010) *Toolkit for Ethical Reflection and Communication*. Workpackage 4: Ethical and Social Impacts. ObservatoryNano: European Observatory for Science-Based and Economic Expert Analysis of Nanotechnologies. ObservatoryNano: Larsim, France. http://www.observatorynano.eu/project/filesystem/files/Toolkit%20full%20final%20 22Jun2010.pdf. Accessed 5 July 2010.

People Science and Policy Ltd (2002) *Biobank UK: a Question of Trust: A Consultation Exploring and Addressing Questions of Public Trust*. Report Prepared for The Medical Research Council and The Wellcome Trust, March. London: People Science

and Policy Ltd. http://www.ukbiobank.ac.uk/docs/consultation.pdf. Accessed 12 August 2010.

Peters, H. P., Brossard, D., de Cheveigné, S., Dunwoody, S., Kallfass, M., Miller, S. and Tsuchida, S. (2008) 'Science-media interface: it's time to reconsider, *Science Communication*, 30: 266–276.

Petersen, A. (1994) 'Community development in health promotion: empowerment or regulation?', *Australian Journal of Public Health*, 18, 2: 213–217.

Petersen, A. (1999) 'The portrayal of research into genetic-based differences of sex and sexual orientation: a study of "popular" science journals, 1980 to 1997', *Journal of Communication Inquiry*, 23, 2: 163–182.

Petersen, A. (2001) 'Biofantasies: Genetics and medicine in the print news media', *Social Science and Medicine*, 52, 1255–1268.

Petersen, A. (2002) 'Replicating our bodies, losing our selves: News media portrayals of human cloning in the wake of Dolly', *Body & Society*, 8, 4: 71–90.

Petersen, A. (2005) 'Securing our genetic health: engendering trust in UK Biobank', *Sociology of Health and Illness*, 27, 2: 271–292.

Petersen, A. (2006) 'The genetic conception of health: is it as radical as claimed?', *Health*, 10, 4: 481–500.

Petersen, A. (2007a) 'Biobanks' "engagements": engendering trust or engineering consent?', *Genomics, Society and Policy*, 3, 1: 31–43.

Petersen, A. (2007b) *The Body in Question*. Routledge: London and New York.

Petersen, A. (2007c) 'Is the new genetics eugenic?: interpreting the past, envisioning the future', *New Formations*, 60: 79–88. (Special issue on eugenics)

Petersen, A. (2009) 'The ethics of expectations: biobanks and the promise of personalised medicine', *Monash Bioethics Review*, 28, 1: 5.1–5.12.

Petersen, A., Andersen, A. and Allan, S. (2005) 'Science fiction/science fact: medical genetics in news stories', *New Genetics and Society*, 24, 3: 337–353.

Petersen, A., Bleakley, A., Brömer, R. and Marshall, R. (2008) The medical humanities today: humane health care or tool of governance?, *Journal of Medical Humanities*, 29: 1–4.

Petersen, A. and Bunton, R. (2002) *The New Genetics and the Public's Health*. Routledge: London and New York.

Petersen, A., Davis, M., Lindsay, J. and Fraser, S. (2010) 'Healthy living and citizenship: an overview', *Critical Public Health*, 20, 4: 1–10.

Petersen, A. and Lupton, D. (1996) *The New Public Health: Health and Self in the Age of Risk*. Allen & Unwin: Sydney and Sage: London.

Petersen, A. and Seear, K. (2009) 'In search of immortality: the political economy of anti-aging medicine', *Medicine Studies*, 1, 3: 267–279.

Petersen, A. and Seear, K. (forthcoming) 'Technologies of hope: techniques of the online advertising of stem cell treatments', *New Genetics and Society*.

Pirmohamed, M., James, S., Meakin, S., Green, C., Scott, A. K., Walley, et al. (2004) 'Adverse drug reactions as cause of admission to hospital: prospective analysis of 18, 820 patients', *British Medical Journal*, 329: 15–19.

Quine, M. S. (1996) *Population Politics in Twentieth-Century Europe*. Routledge: London.

Rabinow, P. (1992) 'Artificiality and enlightenment: from sociobiology to biosociality', in J. Crary and S. Kwinter (eds) *Incorporations*. Urzone: New York.

Rabinow, P. (1999) *French DNA: Trouble in Purgatory*. University of Chicago Press: Chicago and London.

Rapp, R. (1988) 'Chromosomes and communication: the discourse of genetic counselling', *Medical Anthropology Quarterly*, 2: 143–157.

Rawlinson, M. C. (2008) 'Introduction', *The International Journal of Feminist Approaches to Bioethics*, 1, 1: 1–6.

Reardon, J. (2005) *Race to the Finish: Identity and Governance in an Age of Genomics*. Princeton University Press: Princeton and Oxford.

Rehmann-Sutter, C. (2009) 'Why non-directiveness is insufficient: ethics of genetic decision making and a model of agency', *Medicine Studies*, 1, 2: 113–129.

Resta, R.G. (1997) 'Eugenics and nondirectiveness in genetic counselling', *Journal of Gentric Counseling*, 6: 255–258.

Rifkin J. (1998) *The Biotech Century: How Genetic Commerce Will Change the World*. Phoenix: London.

Rose, H. (2001a) 'Gendered genetics in Iceland', *New Genetics and Society*, 20, 2: 119–138.

Rose, H. (2001b) *The Commodification of Bioinformation: The Icelandic Health Sector Database*. The Wellcome Trust: London.

Rose, N. (2007) *Politics of Life Itself: Biomedicine, Power and Subjectivity in the Twenty-First Century*. Princeton University Press: Princeton, NJ and Oxford.

Rose, N. and Novas, C. (2005) 'Biological citizenship', in A. Ong and S. J. Collier (eds) *Global Assemblages: Technology, Politics and Ethics as Anthropological Problems*. Blackwell: London.

Rothman, D. J. (1991) *Strangers at the Bedside: A History of How Law and Bioethics Transformed Medical Decision Making*. Aldine de Gruyter: New York.

Rout, M. (2009) 'Doctor hounded by Vioxx drug reps', *The Australian*, 15 April, p. 3.

Salant, T. and Santry, H. P. (2006) 'Internet marketing of bariatric surgery: contemporary trends in the medicalization of obesity', *Social Science and Medicine*, 62, 10: 2445–2457.

Salter, B. (2007) 'Bioethics, politics and the moral economy of human embryonic stem cell science: the case of the European Union's Sixth Framework Programme', *New Genetics and Society*, 26, 3: 269–288.

Salter, B. and Jones, M. (2005) 'Biobanks and bioethics: the politics of legitimation', *Journal of European Public Policy*, 12, 4: 710–732.

Salter, B. and Salter, C. (2007) 'Bioethics and the global moral economy: the cultural politics of human embryonic stem cell science', *Science, Technology and Human Values*, 32: 554–581.

Sample, I. (2004) 'Nanotechnology poses threat to health, says scientists', guardian.co.uk. 30 July. http://www.guardian.co.uk/science/2004/jul/30/sciencenews.nanotechnology. Accessed 4 July 2010.

Samuel, G. (2010) 'The perils of creating synthetic life', *BIONEWS*, 559, 24 May. http://www.bionews.org.uk/page_61626.asp?dinfo=XdPLnoBGA55enlspQ0T DjuBO. Accessed 6 August 2010.

Schot, J. and Rip, A. (1997) 'The past and future of constructive technology assessment', *Technological Forecasting and Social Change*, 54, 2: 251–268.

Schutz, A. (1972) *The Phenomenology of the Social World*. Northwestern University Press: Evanston, IL.

Scully, J. L. (2008) *Disability Bioethics: Moral Bodies, Moral Difference*. Rowman and Littlefield Publishers, Inc.: Lanham, MD.

Secko, D. M., Preto, N., Niemeyer, S. and Burgess, M. M. (2009) 'Informed consent in biobank research: a deliberative approach to the debate', *Social Science and Medicine*, 68: 781–789.

Selgelid, M. J. (2005) 'Ethics and infectious disease', *Bioethics*, 19, 3: 272–289.

Shakespeare, T. (1998) 'Choices and rights: eugenics, genetics and disability equality', *Disability and Society*, 13, 5: 665–681.

Sharp, A. (2009) 'Drug industry set for lower margins', *The Age* (Business Day), 14 April, p. 3.

Shelley, M. (1985; orig. 1818) *Frankenstein or The Modern Prometheus*. Penguin: Strand, London.

Sherwin, S. (2008) 'Wither bioethics?: how feminism can help reorient bioethics', *The International Journal of Feminist Approaches to Bioethics*, 1, 1: 7–26.

Silver, L. M. (1998) *Remaking Eden: Cloning, Genetic Engineering and the Future of Human Kind?* Weidenfeld and Nicolson: London.

Singer, P. A. and Viens, A. M. (2008) *The Cambridge Textbook of Bioethics*. Cambridge University Press: Cambridge.

Skene, L. (2009) 'Should women be paid for donating their eggs for human embryo research?', *Monash Bioethics Review*, 28, 4: 29.1–4.

Sparrow, R. (2007) 'Revolutionary and familiar, inevitable and precarious: rhetorical contradictions in enthusiasm for nanotechnology', *Nanoethics*, 1: 57–68.

Stein, R. (2010) 'Americans go shopping for old genes', *theage.com.au*, May 12. http://www.theage.com.au/world/americans-go-shopping-for-old-genes-20100511-uunb.html. Accessed 12 May 2010.

Stock, G. and Campbell, J. (eds) (2000) *Engineering the Human Germline: An Exploration of the Science and Ethics of Altering the Genes We Pass to Our Children*. Oxford University Press: New York and Oxford.

Stone, J. (2009) 'Genetic data company deCODE has filed for bankruptcy', *BIONEWS*, 27 November. http://www.bionews.org.uk/page_51447.asp?iruid=294. Accessed 30 November 2009.

Stone, J. (2010) 'Biotech gets final go-ahead for landmark stem cell trial', *BIONEWS*, 15 February. http://www.bionews.org.uk/page_54571.asp?iruid=294. Accessed 19 February 2010.

Suda, E. (2008) 'Causes of communication gaps between science and society: discussion based on community engagement in the haplotype mapping project in Japan', *Eubios Journal of Asian and International Bioethics*, 18, 6: 162–164.

Swierstra, T. and Rip, A. (2007) 'Nano-ethics as NEST ethics: patterns of moral argumentation about new and emerging science and technology', *Nanoethics*, 1: 3–20.

Takala, T. and Häyry, M. (2007) 'Benefitting from past wrongdoing, human embryonic stem cell lines, and the fragility of the German legal position', *Bioethics*, 21, 3: 150–159.

The President's Council on Bioethics (2002) *Human Cloning and Human Dignity: An Ethical Inquiry*. July 2002. The President's Council on Bioethics: Washington, D.C. www.bioethics.gov.

The Royal Society (2005) *Personalised Medicines: Hopes and Realities*. The Royal Society: London.

The Royal Society and The Royal Academy of Engineering (2004) *Nanoscience and Nanotechnologies: Opportunities and Uncertainties*. The Royal Society: London.

The Wellcome Trust & MRC (2002) *Protocol for the UK Biobank: A Study of Genes, Environment and Health*. The Wellcome Trust: London.

Thomson, J. A., Itskovitz-Eldor, J., Shapiro, S. S., Waknitz, M. A., Swiergiel, J. J., Marshall, V. S. and Jones, J. M. (1998) 'Embryonic stem cell lines derived from human blastocysts', *Science*, 282: 1145–1147.

Tong, R. (2001) 'Is a global bioethics possible as well as desirable?: A millennial feminist response', in G. Anderson and A. Santos (eds) *Globalizing Feminist Bioethics: Crosscultural Perspectives*. Westview Press: Boulder, CO.

Triendl, R. and Gottweis, H. (2008) 'Governance by stealth: large-scale pharmacogenomics and biobanking in Japan', in H. Gottweis and A. Petersen (eds) *Biobanks: Governance in Comparative Perspective*. Routledge: London and New York.

Turner, L. (2009) 'Anthropological and sociological critiques of bioethics', *Bioethical Inquiry*, 6: 83–98.

Tutton, R. and Corrigan, O. (2004) *Genetic Databases: Socio-Ethical Issues in the Collection and Use of DNA*. Routledge: London and New York.

Urner, M. (2010) 'Is stem cell research being sabotaged by a "clique"?', *BIONEWS*, 8 February. http://www.bionews.org.uk/page_54229.asp?iruid=294. Accessed 12 February 2010.

van de Poel, I. (2008) 'How should we do nanoethics?: a network approach for discerning ethical issues in nanotechology', *NanoEthics*, 2: 25–38.

Vehmas, S. (1999) 'Discriminative assumptions of utilitarian bioethics regarding individuals with intellectual disabilities', *Disability and Society*, 14, 1: 37–52.

von Braun, J. (2008) *Food and Financial Crises: Implications for Agriculture and the Poor*. Food Policy Report. International Food Policy Research Institute: Washington, DC.

von Schomberg, R. and Davies, S. (eds) (2010) *Understanding Public Debate on Nanotechnologies: Options for Framing Public Policy*. Publications Office of the European Commission: Luxembourg.

Walker, A. P. (1998) 'The practice of genetic counselling', in D. L. Baker, J. L. Schuette and W. R. Uhlmann (eds) *A Guide to Generic Counseling*. Wiley-Liss: New York.

Walker, M. U. (2009) 'Groningen naturalism in bioethics', in H. Lindemann, M. Verkerk and M. U. Walker (eds) (2009) *Naturalized Bioethics: Toward Responsible Knowing and Practice*. Cambridge University Press: Cambridge.

Walters, K. (2009) 'Biotech: the treatment regime of chronic illness is set to change', *BRW*, March 19–25, p. 23.

Waters, R. (2010) 'Fate therapeutics, MIT scientist get stem cell patent', *Bloomberg Business Week*, February 4. http://www.businessweek.com/news/2010–02–04/fate-therapeutics-mit-scientist-get-stem-cell-patent-update1-.html. Accessed 2 March 2010.

Weckert, J. (2007) 'Editorial', *Nanoethics*, 1: 1–2.

West, C. and Zimmerman, D. H. (1991) 'Doing gender', in J. Lorber and S. A. Farrell (eds) *The Social Construction of Gender*. Sage: Newbury Park, CA.

Williams, C. & Wainwright, S. P. (2010) 'Sociological reflections on ethics, embryonic stem cells and translational research', in B. J. Capps and A. V. Campbell (eds) *Contested Cells: Global Perspectives on the Stem Cell Debate*. Imperial College Press: London.

Williams-Jones, B. (2006) 'Be ready against cancer, now: direct-to-consumer advertising for genetic testing', *New Genetics and Society*, 25, 1: 89–107.

Wilmut, I., Schnieke, A. E., McWhir, J., Kind, A. J. and Campbell, K. A. (1997) 'Viable offspring derived from foetal and adult mammalian cells', *Nature*, 385: 810–813.

Wilsdon, J. and Willis, R. (2004) *See-Through Science: Why Public Engagement Needs to Move Upstream*. Demos: London.

Wolf, S. M. (1996) 'Introduction: gender and feminism in bioethics', in S. M. Wolf (ed) *Feminism and Bioethics: Beyond Reproduction*. Oxford University Press: New York and Oxford.

Wolpe, P. R. (1998) 'The triumph of autonomy in American bioethics: a sociological view', in R. DeVries and J. Subedi (eds) *Bioethics and Society: Constructing the Ethical Enterprise*. Prentice-Hall: Upper Saddle River, NJ.

Wood, S., Geldart, A. and Jones, R. (2008) 'Crystallizing the nanotechnology debate', *Technology Analysis and Strategic Management*, 20, 1: 13–27.

Wood, S., Jones, R. and Geldart, A. (2007) *Nanotechnology: From the Science to the Social. The Social, Ethical and Economic Aspects of the Debate*. Economic and Social Research Council. .http://www.esrcsocietytoday.ac.uk/ESRCInfo-Centre/Images/ESRC_Nano07_tcm6–18918.pdf. Accessed 4 June 2010.

Woodrow Wilson International Centre for Scholars. (2010) *The Project on Emerging Technologies*. http://www.nanotechproject.org/inventories/consumer/. Accessed 4 June 2010.

World Health Organization (2002) *Genomics and World Health*. Report of the Advisory Committee on Health Research. World Health Organization: Geneva. http://whqlibdoc.who.int/hq/2002/a74580.pdf. Accessed 18 December 2009.

Wynne, B. (2006) 'Public engagement as a means of restoring public trust in science—hitting the notes, but missing the music?', *Community Genetics*, 9: 211–220.

Index

Locators for headings which also have subheadings refer to general aspects of that topic

*For Product Safety Concerns and Information please contact
our EU representative GPSR@taylorandfrancis.com Taylor & Francis
Verlag GmbH, Kaufingerstraße 24, 80331 München, Germany*

T - #0122 - 160425 - C0 - 229/152/10 - PB - 9780415851398 - Gloss Lamination